Robert Barnwell Roosevelt

Superior fishing; or, the Striped Bass, Trout, and Black Bass of the Northern States

Robert Barnwell Roosevelt

Superior fishing; or, the Striped Bass, Trout, and Black Bass of the Northern States

ISBN/EAN: 9783337143886

Printed in Europe, USA, Canada, Australia, Japan

Cover: Foto ©berggeist007 / pixelio.de

More available books at **www.hansebooks.com**

SUPERIOR FISHING;

OR,

THE STRIPED BASS, TROUT, AND BLACK BASS

Of the Northern States.

EMBRACING FULL DIRECTIONS FOR DRESSING ARTIFICIAL FLIES
WITH THE FEATHERS OF AMERICAN BIRDS; AN
ACCOUNT OF A SPORTING VISIT TO LAKE
SUPERIOR, ETC., ETC., ETC.

By ROBERT B. ROOSEVELT,

AUTHOR OF "THE GAME FISH OF NORTH AMERICA," "THE GAME BIRDS OF
OUR NORTHERN COASTS," ETC.

NEW YORK:

CARLETON, PUBLISHER, 413 BROADWAY.
M DCCC LXV.

R. CRAIGHEAD, PRINTER,
Caxton Building, Centre St., N. Y.

INDEX.

SUPERIOR FISHING.

GENERAL REMARKS.

ALTHOUGH the shores of our northern coasts, both along the Pacific and Atlantic Oceans, abound in numberless varieties of the finny tribe, and myriads of striped bass, cod, mackerel, tautog, herring, shad and blue-fish in the Northern States, and salmon, sea-trout, and capelin in the British Provinces, visit us in their season ; the inland States, with the reservation of certain restricted localities, produce few varieties, and with a single exception, inferior kinds of fish. Throughout that vast region west of Pennsylvania, bordering on the great lakes, and stretching westward to the Rocky Mountains and northward to the Canadian boundary, as well as the centre of British America not communicating immediately with the sea or the immense bays of the Arctic Territory, there can be found but one, or at the most two kinds of fish that are worthy of the attention of the epicure or the sportsman. It is true that savage pickerel, immense mascallonge, and gigantic cat-fish lie in wait amid long weeds, and embedded in deep mud, a terror to their smaller brethren and a prize to the unrefined fisherman who looks to the profit to be derived from their heavy carcases; and that other coarse and ill-shapen creatures are taken in the net; but the only fishes that the true angler can regard as objects of sport are the pike-perch, and the black bass.

The pike-perch, which is variously termed the pickerel, pike of the lakes, glass-eye, big-eyed pike, and pickering, is taken in immense numbers in Lakes Erie and Huron, was formerly numerous in the Ohio, and inhabits to a greater or less degree the ponds or sluggish waters of that section. It is a savage fish, biting voraciously at bait or trolling-tackle, and where better fish are scarce, is regarded as a piscatory delicacy; but its play is weak and dull, and as it is taken with strong tackle, its capture requires neither the skill nor experience that lend the principal charm to angling; and by comparison with sea-fish, its flavor is coarse.

Captured mainly with the all-devouring net, it is salted and packed for winter use as our cod or mackerel are preserved, and constitutes at Sandusky and some other places an important object of commerce.

The black bass, a fish that, from its abundance in their country, Americans may claim as peculiarly their own; a fish that is inferior only to the salmon and trout, if even to the latter; that requires the best of tackle and skill in its inveiglement, and exhibits courage and game qualities of the highest order—

fairly swarms in the upper central portion of North America.

In all the lakes, large and small, that dimple the rugged surface of Canada; in the sheets of pure water embosomed in the gentle swells of the western prairies; in those inland seas that are enveloped by our extensive territory; and in the numerous rivers of the west—the black bass is found by his ardent admirers.

From the confines of Labrador, throughout the Canadas, in British America, the Western States, and far beyond the Mississippi, there is scarcely a stretch of water, whether it be the rapids of the St. Lawrence, the sluggish bays of Lakes Ontario and Erie, the cold depths of Huron and Superior, or the lakelets of the interior, that does not abound with this splendid fish.

In dull weedy bays he becomes lazy, ugly, and ill-flavored; but in cold or rapid water, or upon stony bottom, he acquires a vigor of body and excellence of flavor that place him in the first rank of piscatorial prizes.

Although not abundant, if even indigenous in the Middle States, he has been extensively introduced; and finding many of the clear, transparent, rocky, eastern ponds admirably adapted to his health and propagation, he is populating waters that have heretofore produced little besides perch and sun-fish. By a fortunate provision of nature, most ponds that are not suited to trout are favorable to black bass; and being a hardy fish, able to endure long journeys, he is readily transported from place to place. The time will soon come when the worthless yellow perch will be supplanted by his noble congener.

He has been imported even into that semi-detached point of New England, Cape Cod, and thrives wonderfully in Lake Mahopac, adding much to the attractions of that favorite watering-place of fashion-jaded New Yorkers, and is being generally distributed among his eastern friends. If not exposed to a hot sun, he may be carried a long distance out of water, and will often revive when apparently the last spark of vitality is extinct. But his natural home is north and west of the Middle and Eastern States; there his name is legion, his fame deservedly great, and he may be almost said to be the one game fish.

It is true that among epicures the famous white fish of Lakes Huron and Superior, which is also found in a more flabby condition in Erie and Ontario, ranks before either the black bass or the pike-perch; but as he is deceived by neither decoy nor bait, he is not worthy of the fisherman's regard. To be tasted in perfection, the white fish must be eaten fresh from the rapids of Lake Superior, where, lying in the eddy below some immovable rock, he is taken by the sharp-eyed Indian in the long-handled net from out the foaming water, brought immediately to land, cooked and placed steaming hot upon the table before he has lost the delicious freshness of his native element.

The black bass, however, is in the west what the

trout is in our eastern brooks—the principal source of the angler's enjoyment.

The rivers that empty into Hudson's Bay are ascended by the migratory salmon, but from their peculiar character do not furnish fly-fishing except for trout. The latter are found in Lake Superior and the streams that empty into it, in the tributaries of the Upper Mississippi, and in the brooks of the Alleghany and Rocky Mountains; but are not generally distributed through the weedy streams of the Western States.

The flat expanse of Ohio is not favorable to the existence of that lover of the noisy brook and tumbling torrent; and streams flowing through marl deposits are supposed not to furnish proper food; so that the beauty that we in our eastern homes entice from every stream or brooklet from Maine to Pennsylvania, is found rarely, if at all, in Illinois, Indiana, Ohio, western Kentucky, and southern Wisconsin; but in the cool depths of Lake Superior and its amber-hued tributaries he absolutely swarms.

In the Upper Mississippi there are black bass and mascallonge; in the brooks that, rising amid the hills of that region, swell its current, there are trout; in neighboring lakes black bass and perch abound; among the Rocky Mountains are found several species of trout; and in the waters of Oregon and California salmon are plentiful.

Although the largest trout in the United States are taken in Maine, in the Umbagog region, the greatest number and the most vigorous are found in

Lake Superior, where fish of two pounds weight can be captured to the heart's content. The fish of Maine are of rich and strong color, while those of Lake Superior have the bright sides and delicate tints of the sea-trout. All brook trout, however— the genuine *salmo fontinalis*—have the peculiar bright vermilion specks that distinguish them from kindred species, and these are distinctly visible upon the silver sides of the fish of Lake Superior.

The innumerable rivers of the State of Maine are interwoven together in such a manner that the fisherman, urging his silent canoe with dripping paddle or stout pole, gliding beneath the arching boughs that shade in gloom the narrow stream, or pushing boldly into the open lakes, can pass from one region of waters to another, and, making short portages, explore in a continuous trip rivers that run north, east, and west. To the true sportsman, armed with pliant rod and feathered hook for the seduction of the merry trout, and trusty rifle loaded with heavy ball for the destruction of the lordly moose, nothing surpasses the intense enjoyment of wandering amid the forest wilds from river to river, threading the uninhabited groves, or following the unknown and unnamed stream, and leaving to whim or chance, or the influence of luck, to determine his final destination. Alone with his single guide he is content; accompanied by a friend, still better pleased; in a party of associates perfectly happy; blessed by the society of ladies—real ladies and true wood nymphs —he is in Elysium.

Or, he may coast the shores of our western lakes, where the bright sun sparkles on the rippling surface, and only seek the shade upon the land to avoid its heat; there he may kill the black bass, the mascallonge, and in Lake Superior the trout; fleeing from the approaching storm to some sheltered nook, he partakes the inland ocean's varying moods, passing the days upon its surface and the nights amid the neighboring forests; stopping occasionally to use the light shot-gun and kill a few woodcock or partridges, and now and then slaying a duck upon the route.

In the wide world there is no other country so propitious to the fisherman as the northern part of North America; it furnishes every variety of sport, from the delicate refined fishing of the transparent ponds and over-fished trout-preserves of Long Island, to the coarser and easier sport of killing with large flies and heavy rods the countless hosts of Maine, the Labrador coast, or Lake Superior; from the casting the menhaden bait into the boisterous ocean for striped bass, to the trolling amid the Thousand Isles of the St. Lawrence for the ugly and powerful mascallonge; from the capture of the noble salmon to that of the spirited black bass. In fact, there is so much and so good fishing everywhere, that it is difficult to give a preference or lay out any specific directions. You may go by railroad to Cape Vincent, and thence by steamboat to Clayton or Alexandria bay, and fish the St. Lawrence; or take the ocean steamer from Boston to Eastport, and thence to Calais, and explore the St. Croix River for land-locked salmon; or continue on to St. John, and by rail-road and stage or steamer to the Nipisiquit, and kill the true salmon—*salmo salar*—king of fish; or you may take the railroad from Boston to Bethel and cross by stage into the Umbagog region of Maine, and visit its innumerable lakes with unpronounceable names, or may embark on the steamboat at Cleveland, and wake up, after two days' tranquil voyage, at the Sault Ste. Marie, the outlet of Lake Superior; or you may stop anywhere on any of these routes, even out in the ocean, on the way to New Brunswick, if you please, where there are pollock or haddock, and have good fishing. There is excellent fishing close to New York city, and better still the farther you recede from it.

It is true the fisherman will not find those refined comforts that the more cultivated and densely peopled districts of Europe afford; but he will receive a hearty welcome and wholesome entertainment at the country tavern or the farmer's house. If, however, he have youth and tolerable hardihood, he should look for no such reception; but, carrying his canvas-home, enjoy the luxury of unrestrained independence, kill and cook his own dinner, and sleep in the pure air of the wilderness. He will have to surrender a few necessaries that habit has made so, but he will be repaid a thousandfold by increased happiness and improved health; he will not have servants to wait on him, nor desserts or wines to pamper him; but he will have his guide to instruct, and abundant food to support him. He will acquire an insight into the mystery of woodcraft, and learn a few of its wonders and delights; he will come to rely upon his own stout muscles and sharp eyes, and return to the city a renovated being. Or, if he have sufficient enthusiasm and high courage, he may cast aside all trammels, and taking his rifle or rod, salt pork, and hard bread, strike off into the trackless forest with no covering to shield him from the rain or sun, no floating thing of beauty to bear him in its bosom over the water, no store of provisions to fall back upon if fish do not rise and the bullet flies astray; but bearing bravely up against heat and weariness, sleeping, amid the rain and storm, wrapped in the heavy coat, catching or killing game sufficient for daily food, or going hungry till better luck shall interpose. This, indeed, is manhood; and our country, with its vast solitudes, its unbroken forests, its network of water-courses, its endless chains of lakes, its vast mountains and limitless prairies, offers inducements for such a life that no other land possesses.

As pretty full instructions have been given in the *Game Fish of North America* to aid the learner in commencing his experiences of camp life, the reader who desires such information is referred to that work; but whether he shall go into the solitary wilderness, away from man and human habitation, or can only tear himself from business for a few hours for a flying visit to some quiet preserve near the bustling city, he should never forget that he is a sportsman,

and owes the duties of moderation, humanity, patience, and kindness under all circumstances; that
he cannot slaughter or poach; and that, from his pro-
fession, he should ever be a gentleman. He should
never forget the words of that most amiable of our
fraternity—the splendid shot, the skilful angler, the
genial companion, and the graceful writer, now long
since gathered to his final resting-place—who was
known to the public under the name of J. Cypress,
Jr.:

"No genuine piscator ever tabernacled at Fire-
place or Stump-pond who could not exhibit proofs
of great natural delicacy and strength of apprehen-
sion—I mean of things in general, including fish.
But the *vis vivida animi*, the *os magna sonans*, the
manus mentis, the divine rapture of the seduction
of a trout, how few have known the apotheosis!
The creative power of genius can make a feather-
fly live, and move, and have being; and a wisely
stricken fish gives up the ghost in transports. That
puts me in mind of a story of Ned Locus. Ned
swears that he once threw a fly so far and delicately
and suspendedly, that just as it was dropping upon
the water, after lying a moment in the scarcely
moving air as though it knew no law of gravity,
it actually took life and wings, and would have
flown away but that an old four-pounder, seeing it
start, sprang and jumped at it full a foot out of his
element, and changed the course of the insect's travel from the upper air to the bottom of his throat.
That is one of Ned's, and I do not guarantee it, but

such a thing might be. Insects are called into being in a variety of mysterious ways, as all the world knows; for instance, the animalcula that appear in the neighborhood of departed horses; and, as Ned says, if death can create life, what is the reason a smart man can't? Good fishermen are generally great lawyers; *ecce signa*, Patrick Henry and Daniel Webster. I have known this rule, however, to have exceptions. But the true sportsman is always at least a man of genius and an honest man. I have either read or heard some one say, and I am sure it is the fact, that there never was an instance of a sincere lover of a dog, gun, and rod being sent to bridewell or penitentiary . . . . If I were governor and knew a case, I would exert the pardoning power without making any inquiry. I should determine without waiting to hear a single fact that the man was convicted by means of perjury. There is a plain reason for all this. A genuine sportsman must possess a combination of virtues which will fill him so full that no room can be left for sin to squeeze in. He must be an early riser— to be which is the beginning of all virtue—ambitious, temperate, prudent, patient of toil, fatigue, and disappointment; courageous, watchful, intent upon his business; always ready, confident, cool; kind to his dog, civil to the girls, and courteous to his brother sportsmen."

To constitute a sportsman, therefore, it is not sufficient merely to be able to catch fish; although a very important element in the angler's composition, it is not all that is required, nor will it alone entitle him to full fellowship with the fraternity. He must have higher aspirations and nobler gifts; he must look beyond the mere result to the mode of effecting it, regarding, perhaps, the means more than the end. Any unfair trick or mean advantage he must never take, even to fill a vacant creel or empty pocket; he must never slay the crouching bevy, huddled in terror before his pointer's nose; he must never resort to the grapple or the noose, no matter how provokingly the wary trout, lying motionless in the clear water, may disdain his choicest flies; and, when the nature of the fish pursued induces it to accept the imitation, he can use the natural bait, only in extreme cases and at great risk to his reputation. The noblest of fish, the mighty salmon, refuses bait utterly, and only with the most artistic tackle and the greatest skill can he be taken; the trout, which ranks second to the salmon, demands an almost equal perfection of both, and in his true season, the genial days of spring and summer, scorns every allurement but the tempting fly. The black bass prefers the fly, but will take the trolling-spoon, and even bait, at all seasons; whereas the fish of lesser station give a preference to bait, or accept it alone. This order of precedence sufficiently proves what every thorough sportsman will endorse—that bait-fishing, although an art of intricacy and difficulty, is altogether inferior to the science of fly-fishing; and that the man who merely follows it without higher aspiration, and uses a worm equally for the beautiful trout and the hideous cat-fish, cannot claim to be a sportsman. Occasionally there is a person who will use the bait with wonderful ability, and entice the reluctant fish against their will to an unwished-for meal; but he never experiences the higher pleasures of his pursuit—his enjoyment in making a neat and killing fly, his satisfaction at its success, his delight in putting it properly upon the water, and his gratification when with it and his frail tackle he shall have overcome the fierce and stubborn prey. Therefore to his many other qualities, the true sportsman must add a thorough knowledge of fly-fishing, and only can the use of artificial fish or fly, or casting the menhaden bait for bass, be termed SUPERIOR FISHING.

THE FIRST VEIL.

CHAPTER I.

LAKE SUPERIOR.

Don Pedro is descended from one of what we in our young country call the old and highly-respectable families, and having been nurtured amid the refinements and luxuries of life, is one of the most gentlemanly men imaginable. At the public rooms of a hotel, in the halls, on the piazza, in the saloon of a steamboat, he can never pass a lady, though she be a perfect stranger, without in the most deferential manner removing his hat. To this reverence for the fair sex he adds an easy elegance towards his own, that at once commands attention and respect.

Never having taken an active share in the world's affairs, his abilities, which are far above the average, have lain dormant or run to criticising art or committing poetry; and he is rather apt to discuss very small matters with a minuteness and persistency that important ones scarcely merit.

He had travelled Europe, of course, had shot quail and taken trout in Long Island, fired at crocodiles on the Nile and jackals in the desert; and although probably the greatest exposure of his life had been damp sheets at a country inn, and his

severest hardship the finding his claret sour or
being compelled twice in one day to eat of the same
kind of game, he was now seized with a sporting
mania, and determined to rough it in the woods.
An unsafe companion, perhaps, the reader may
think; but it is not always the roughest men who
have the most pluck, nor those accustomed to the
commonest fare who grumble the least when offered
still coarser, and there is truth in the words of worthy Tom Draw: " Give me a raal gentleman, one
as sleeps soft and eats high, and drinks highest
kind, to stand roughing it."

So we discussed matters over a comfortable dinner, with the aid of a couple of bottles of claret,
one of champagne, and a little brandy; and Don
concluded he would as lief eat salt pork as woodcock, and ship biscuit as French rolls. He was
anxious to examine my list of camp articles, and was
quite ready to do away with a large part of them;
but finally determined to leave that matter to me,
holding me strictly responsible for carrying any
unnecessary effeminate luxuries. The discussion
was not a short one, but this happy decision being
arrived at, I was perfectly satisfied.

We met by appointment a few days later at the
Angier House, in that thriving, active town of Cleveland, which seems to be drawing to itself the business of the other cities of Lake Erie, and, cannibal-
like, to be growing fat on their exhausted lives. It
is a thoroughly American city, and, like all our
cities, doubtless has the handsomest street in

the world, for so we were assured by the citizens.

A large part of the trade of Cleveland is with the mines of Lake Superior, and steamers leave almost daily for that region, carrying a miscellaneous assortment of the necessaries of life, and returning laden with copper and iron ore. Not content, however, with this unexciting freight, these vessels propose to carry excursion parties round the lakes, and are all, if their advertisements are to be believed, supplied with brass bands, and every luxury of the season.

In Cleveland we intended to purchase such ardent spirits as we might require, and Don commenced :

"Now as to this question of liquor, I should like to have your views concerning kind and quantity ?"

"Well, I expect we will be in the woods twenty days, and have made my computations on that basis; so we will need a case of liquor, and as you prefer brandy, brandy let it be."

"No, no ; by no means," responded Don ; " do not let my predilections influence you ; besides, a dozen bottles seems a good deal. If we were gone twenty-four days it would be just a pint a day, or a half-pint apiece—rather severe, considering we expect to rough it."

"You know we have to give the men some occasionally, and then we will meet other parties and have mutual good-luck to drink. It will not be an over-supply, though we can make it less if you say so; I myself drink little when in the woods."

"I believe that," replied Don, ironically; "and considering how well I know you, it was hardly worth while to mention it. But this is a serious question, for we can get nothing drinkable after leaving Cleveland; and if we have to do what you say, do you not think we shall run short? I want plenty of everything, and it would be better to take a dozen and a half, if there is a doubt."

"There is no doubt; but if——"

"If you say there is no doubt, that is sufficient; but I am surprised you should give the men expensive brandy, when they would probably prefer a coarser article."

"Of course, we will take a common whiskey for the men; but occasionally while using the flask ourselves we will naturally pass it to them."

"Ah, yes; I understand. But, really, I am not satisfied it should be all brandy; you must not expect to have the same comforts you would in the city, and if you will take my advice, you will have at least part whiskey."

"But you prefer brandy, and one is as easy to carry as the other."

"Really, now, you must not consult my wishes; in fact, although I admit a slight preference for brandy, many persons prefer whiskey. Before you decide, it would be well to examine the matter thoroughly; and as we are now at the store, you must make up your mind promptly."

This conversation had taken place as we were

2

walking from the hotel to an establishment that had been recommended to us.

"Remember," continued Don, "you must act for the joint interest, and there are several points well worth considering. In the first place, whiskey is much cheaper; then it is probably purer than the brandy you buy here; if a bottle should be broken the loss is less——"

"Certainly; if you would be equally content, I should arrange it differently."

"How often must I tell you not to consider me, and I am decidedly pleased at your change of views. Now, putting aside any supposed preference on my part, what proportions would you suggest?"

"Nine of whiskey to three of brandy."

"Ah," gasped Don, losing his breath at the suddenness of this response, "have you given the matter sufficient consideration? You have not even ascertained the price;" and then turning to the clerk, he asked: "How do you sell your best whiskey?"

"Eight dollars a dozen, and brandy two dollars a bottle."

"Nine bottles of whiskey would be six dollars," I calculated aloud, "and six for the brandy, make twelve. Have them packed and delivered on board the *City of Cleveland* promptly at half-past seven, because she leaves at eight."

"But are you satisfied?" cried Don in an agony of horror at such a want of discussion; "have you examined all the bearings of the change? Can it

be packed in time? You know whiskey does not
go as far as brandy. Are you sure you have
enough? Is there no question about that being
the best proportion? Would you not prefer all
whiskey? In case of sickness, may we not need
more brandy? What is the best mode of packing
it? Is it sure to be at the boat punctually?"

"That is the clerk's affair; if it is there it will be
paid for, and if not it won't. Let's look at the
town; come," and I dragged him off just in time to
avoid a dozen new propositions, and as many unan-
swerable questions, leaving the clerk, bottle in hand,
looking the image of despair at the avalanche of
inquiries that had burst upon him.

After strolling about for several hours we reached
the boat, and found the case of liquor waiting for
us, and proceeded to select our stateroom. This
matter rose at once to a serious question in Don's
eyes. I resolved to leave it entirely to him, confident that his elegant manner would impress the
steward. He at once devoted his entire attention
to it, flitting from place to place in the forward and
after cabins with the steward at his side, pointing
out defects here, suggesting changes there, popping
in and out of doors, describing his foreign experiences and the prime necessity of comfortable quarters, turning down the sheets, peering into cracks,
feeling the pillows, casting a suspicious eye upon
blankets, dissatisfied with all, and finally resolved
to take one which could not be examined at the
time for want of the key, but which the steward,

who had been a respectful and sympathetic listener, assured him had none of the defects he had pointed out.

The immaculate stateroom was engaged, the boat pushed off, the key was obtained, and lo and behold! if it had none of these specified defects, it had another—one of the wooden supports, a huge beam eighteen inches broad, passed directly up through the foot of both the berths, reducing them to four feet six inches in length. When Don made this discovery his face was a study for his friends the artists; anger could not do justice to the occasion; despair, bewilderment, horror, astonishment, seemed blended, with a lurking suspicion that the sympathetic steward had been making game of him. He rushed to the office, could find nothing of the steward, but was informed that all the other staterooms were engaged.

However, after supper, the officials relented and gave us another room, enjoying mightily their joke, as I always believed it to be, although Don never could be brought to admit that they could by any possibility have dared to make fun of him, and insisted it was a blunder of that "stupid steward."

We reached Detroit by five o'clock of the following morning, and as the boat for some wise reason remained there till two in the afternoon, we strolled round the city. It is a promising place, and has the finest street in the world, so the citizens assured us, called Jefferson Avenue. The market was well supplied with fish, and among them sturgeon, cut into slabs of yellow, flabby flesh ; pale Mackinaw salmon, and darker ones from Lake Superior; white fish, the best of which were sold for six cents a pound; lake mullet, black and white bass, yellow and white perch, sun-fish, northern pickerel, suckers, pike-perch, cat-fish, and lake shad or lake sheepshead, called in French *Bossu*, or humpback—a very appropriate appellation. These fish had been for the most part taken in nets; but black bass are captured abundantly with the rod in the small lakes near Detroit, and in Canada opposite. The principal articles sold in the market, however, were strawberries and hoop-skirts; the latter being so numerous that Don remarked incidentally that the inhabitants absolutely skirt the market. This he evidently intended as a joke.

A few miles beyond Detroit is situated its pretentious rival, Port Huron, which is also a flourishing town, and has the handsomest street in the world; and opposite Port Huron are Sarnia and Point Edwards, the termini of the Grand Trunk and the Great Western railroads of Canada. We touched at Point Edwards at about eleven o'clock in the evening.

America is a great place; the people are upright, virtuous, honest, enterprising, energetic, brave, intelligent, charitable and public spirited; they are the finest race of men and the most beautiful and cultivated women in the world, but they do not know how to dine. To gobble down one's victuals, regardless of digestion or decency, is not eating like Chris-

tians but feeding like animals; to thrust one's fork
or spoon into the dish appropriated to holding food
for all, is uncleanly and offensive ; to eat peas with
a knife is bad enough, but to use it immediately
afterwards to cut butter from the butter-plate is
absolutely disgusting.   No one who does these
things is either a lady or a gentlemen; and no one
who cannot keep his arms at his side while cutting
his meat is fit to eat at a public table.

There was one gentleman, as he would claim to
be considered, who sat near us, who, although he
had a proper silver fork, endeavored religiously to
eat his peas on a knife that happened to have a small
point.   This operation, always difficult and dangerous, became, from the formation of the blade, almost
impossible ; the peas rolled off at every attempt, and
the unfortunate rarely succeeded in carrying to his
mouth more than one at a time, till finally reduced
to despair, he seized a table-spoon, and with it devoured them in great mouthfuls.

The dinner was quite a lively scene ; the ladies,
although there was plenty of room, were smuggled
in clandestinely before the gong was sounded, and
the men, dreading the horrors of a second table,
rushed for the remaining chairs, standing behind
and guarding them religiously, but politely waiting
till the ladies were seated.   There was plenty of
food, but each man immediately collected such delicacies as were near him, and he imagined he might
need, and transferred them to his plate or a small
saucer.   There was abundance of time, no one hav-

ing the slightest prospect of occupation after dinner, and yet every man, woman, and child set to work eating as though they expected at any moment to be dragged away and condemned to weeks of starvation.

The waiters, like all Americanized Irishmen, were independent if not insolent, and we overheard the following discourse between one of them and an unhappy wretch who had come in late and could obtain no attendance. The suffering individual began rapping on his plate with the knife till he attracted the notice of a passing waiter:

Waiter.—" Well, what are you making that noise for ?"

Starving Individual.—" I should like to have something to eat."

Waiter.—" Isn't there plenty to eat all round you ?"

Individual.—" But I want some meat."

Waiter.—" Why don't you ask for it, then ? What do you want ?"

Individual.—" What kinds are there ?"

Waiter.—" Why there's beefsteak, to be sure."

Individual.—" I would like to have some beefsteak."

Waiter.—" Why didn't you say so, then, at first ? Give me your plate if you expect me to get it for you."

It was their habit to empty the water left in the glasses back into the pitchers, and when I asked one for a glass of water, he drank out of it himself first, and then handed it to me. On another occasion he helped Don by giving him the tumbler a stranger had just used.

These little peculiarities all round encouraged sociability; you could hardly refuse to know a man when you had drunk out of the same glass and eaten from the same dish with him, and a lady naturally felt at home with a gentleman whose ribs she had been punching for half an hour. The progress of the meal, however, was somewhat checkered, not a few of the guests clamoring for their dessert ere the others had finished their soup. The only explanation of this haste was from the graceful stewardess, who was the redeeming feature of the boat, and who˙ said the waiters were in a hurry so as to have it over as soon as possible. It might aptly be said of the Americans: "They eat to live."

Beyond Lake St. Clair the land on both sides of the river is low, and, especially on the Canadian side, adorned with cultivated farms and dotted with picturesque country houses. A half mile barely separates the two nations; and, in case of war, with our present improved artillery, the intervening river would hardly form an obstacle to mutual destruction, till the once smiling fields and happy homes would be one vast scene of desolation.

Emerging into Lake Huron we began to perceive the effects of the cool water and consequent condensation of the warmer atmosphere; a heavy fog lay upon the surface, at first not higher than our upper deck, but creeping up as the night advanced. On one side a beautiful fog-bow with faint and delicate colors, spanned the sky, while on the other a brilliant ring of sparkling silver surrounded the moon.

The water that was an opaque, milky white at Cleveland, had been growing darker, greener, and clearer, attaining perfect purity ere we reached Lake Superior, and exposing to view objects many feet below its surface.

Having reached Detour, which is a growing place and will soon have the finest street in the world, at eight o'clock at night, and the channel through Lake George being intricate, the captain announced we could proceed no further that evening, and the passengers generally went ashore to explore the country. The land is low around Detour, though there are clusters of pretty islands, and here for the first did we see the rocky northern formation and the evergreen trees.

Lake George, which is at the head of Lake Huron, or more properly a part of it, is shallow and muddy. A channel, narrow and of but twelve feet in depth, has been dredged and marked out with stakes ; it is crooked, and will scarcely admit of two vessels passing abreast. The shoal mud-flats were visible in every direction, and our wheels stirred up the bottom as we passed.

It was with a feeling of relief that we escaped from this lake into the deeper and rapid waters of the river Ste. Marie, whose eddying current and bold shores were a pleasant sight, to our eyes wearied with the sameness of lake travel. We had been three nights and almost three days caged in our floating home, and were delighted at the near approach to our destination. We had not heard the band mentioned in the advertisements, but supplied its place with a crazy piano strummed by amateur performers; we had not partaken of all the luxuries of the season, but had appreciated with sharpened appetites the substantials that were furnished; we had not enjoyed the company of fair excursionists from Cleveland or Detroit, but had formed the acquaintance of one or two kind beings in crinoline; we had not had an exciting trip, but had been transported safely and slowly, and at eight o'clock that morning we reached the Sault Ste. Marie.

A weary waste of waters lay behind; our track lengthening into the dim distance, stretched out to many thousand miles; we had crossed deep streams, had burrowed through high mountains, had darted along broad meadows, had swept across majestic lakes, had ascended mighty rivers; less than a hundred years ago many months would have been expended in completing this same journey; serious difficulties would have had to be overcome and dangers encountered; we had condensed a year of our grandfathers' lives into three days; we had spanned one-half our great continent, fled from the metropolis of civilization to the native haunts of the savage; in fact, gone back from the nineteenth into the eighteenth century. We had been carried by steam upon the track of iron or in the moving palace; in future we were to embark in the voyageur's bateau, and be propelled by oars or sail. Heretofore the unnatural wants of civilized life had been indulged and gratified ; hereafter, the commonest home, the simplest covering, the plainest food, was to be our lot ; hitherto we had been in the land where gold was the talisman that commanded ten thousand slaves ; henceforth we were to trust ourselves to kindly nature and our own capabilities. Glorious were our anticipations from the change. Our vessel, the unromantic *City of Cleveland*, which, from the beginning, had been lumbering along at the moderate rate of ten miles an hour without ever being betrayed into the slightest evidence of enthusiasm, seemed overjoyed at her approaching arrival, and dressed herself in her gala costume of variegated bunting. She whistled merrily to announce to the inhabitants that once more she was to bless their sight, and tried to get up a little extra steam for a final burst. The travellers crowded her decks, the natives collected along shore ; the former waved their handkerchiefs, the latter, probably having no handkerchiefs, swung their hats ; and amid all this excitement we came merrily up to the dock.

The Sault, or Soo, as the name of the village is always pronounced, is not a large place, but proved to be larger than I expected ; our dull plodding eastern people can hardly imagine how rapidly the west is growing in wealth and population ; already our little western brother is claiming to be a man, and if we are not careful will be too much for us some day. This newly planted village, almost at the extreme northwest of American civilization, included an excellent hotel, a dozen stores, and at least a hundred houses and workshops. Already the belles of Illinois, Wisconsin, Iowa, and Minnesota were congregating at it to enjoy its cool temperature and invigorating atmosphere, and ere many years are passed it will be a fashionable watering-place, thronged with the *élite* of western society. Its principal hotel, the Chippewa House, is admirably kept, and doubtless is the pioneer of an infinitely more gorgeous affair.

Don, however, who is rather particular and not much accustomed to the free and easy mode of country life, was somewhat disappointed with our room. It had the great desideratum of plenty of fresh air, for it was of the whole width of the house and had windows back and front, but Don was surprised that people who kept hotels did not acquaint themselves with the other important requisites.

"There, for instance, you observe the water pitcher has a cracked handle. Some time you will undertake to lift it and it will give way, and then there is no telling what it may ruin; the trunk, even, may receive the entire contents."

"But, Don, that is an old crack; it has evidently stood several years, and will doubtless last the few days we are here."

"Not so certain; and just observe that disgusting nick in the wash-basin, it will always look dirty even if it is not."

"Don, you are wrong there; that is a good sign, it proves the basin may nick but won't break."

"Then there is no slop-basin; now what do you suppose we are to do without a slop-basin?"

"Why, throw the slops out of the window, to be sure."

"You would hardly call that decent in New York; and not only may they fall on some passer-by, but the window is too small to permit it conveniently. Just look at this pillow; it is long, to be sure, but not stuffed with half feathers enough; what am I to do with such an apology for a pillow as this?"

"Why, double it up, of course."

"I see," he concluded, in a resigned tone, "you are making a joke of these matters, so we will not pursue the subject; but now that we are on shore fresh from our voyage, I wish to ask seriously your deliberate opinion whether you would advise any one to take the trip just for the pleasure of the journey itself?"

CHAPTER II.

In the northern part of Minnesota is the greatest elevation of what geologists denominate the eastern water-shed of our continent; lying almost exactly in the centre of North America, here the streams that flow to the north, east, and south, find their source. Lake Superior, that adjoins this section on the east, is the chief of those magnificent lakes that empty from one another into the St. Lawrence, and finally wash the coast of Labrador. The Mississippi, taking its rise in the same region and but a few miles away, flows southward with ever increasing volume to the Gulf of Mexico, and then sweeping around Florida and through the Atlantic, rejoins the waters of Lake Superior off Newfoundland; while the Red River of the North, pursuing a contrary course, empties into Hudson's Bay and thence into the Northern Ocean. These waters, starting from little rills and springs scarcely more than a few steps apart, after wandering thousands of miles asunder come together and commingle in the Northern Atlantic Ocean.

Here were the famous Indian portages. One from Lake Superior through Pigeon River, Sturgeon Lake, and Rainy River into the Lake of the Woods, has served to locate the boundary between two great nations, and is the native highway between Hudson's River and Hudson's Bay. Another through Brulé River leads into the head waters of the Mississippi, and thence, by ascending the Missouri, to the rivers that empty into the Pacific Ocean. These portages were traversed year after year by the aboriginal inhabitants, who have left their tracks in the well-worn paths that are still followed by the voyageurs, and are suggestive of easy grades to those who wish to bind our country together by paddle-wheel and railroad track.

Lake Superior, with a surface six hundred feet above, and a bottom three hundred feet below the level of the sea, stretches out in vastness and splendor five hundred miles long by nearly two hundred broad, and holds in its bosom islands that would make respectable kingdoms in the old world. On the southern shore its sandstone rocks are worn by the waves and storms into fantastic shapes, imitative of ancient castles or modern vessels, or are hollowed out into deep caverns; on the north the bolder shore rises into rugged mountains whose face has been seamed by the moving ice-drift of former ages. In the country bordering upon the south are located inexhaustible mines of copper and iron of immense value; and along the northern coast are found agates and precious stones.

A hundred streams pour their contents into the great lake which, from its enormous size and depth, retaining the temperature of winter through the summer months, empties its clear, cold, transparent

waters into the river Ste. Marie. Not producing a
large variety of fish, those that dwell in its bosom
are the finest of their species. The speckled trout,
the Mackinaw salmon, and the black bass are large
and vigorous; sturgeons are plentiful, although
valueless except as an article of food; and the white
fish are the daintiest fresh-water fish in the world.

The forests are mainly composed of the sombre
evergreen trees, relieved frequently by the beautiful
white birch, and along the low lands by a considerable number of other varieties; the shore on the
north is a bold bluff five hundred feet high, but
where it descends to the water it forms occasionally
tracts of fertile interval; on the south the coast is
more level and apparently more sterile. Both shores
are as yet totally uncultivated, and from the severity
of the winters will probably long so remain.

Immediately upon our arrival at the Sault we
made our preparations for a campaign against the
fish, and engaged as guides Joseph Le Sayre, a
Melicete chief, and Alexis Biron, a Canadian half-
breed. Old Joe, as we called him, though he did
not seem over forty, was a fine looking Indian with
an erect graceful shape, and pleasant open countenance; Alexis, though apparently a good man, was
not so prepossessing.

We embarked in a large, stout canoe, and paddling
across the broken water at the foot of the fall, commenced fishing the streams into which the river is
divided by numerous islands near the opposite shore.
A small, brown caddis fly, or, scientifically speaking,

phryganea, covered the water in myriads, was
wafted along in clouds by the wind, and settled upon
the trees and rocks everywhere. Knowing that they
changed from a species of worm on rising to the
surface, we selected clear, calm spots and endeavored to examine the process. It was too rapid for
human eyesight; a spot of transparent water would
be bare one instant, and the next there would be
upon its surface two or or three little creatures dancing about and trying their wings preparatory to a
bolder flight. We never managed to see the larva,
but invariably beheld the perfect fly appear instan-
taneously.

Their number was incalculable; living ones filled
the air, were blown along like moving sand, were
carried into our faces so that we could scarcely face
the wind, and settled upon our boat; dead ones
covered the water in all directions, were devoured
by the fish, especially the lake herring, and were collected by the current in masses resembling sea-
weed. They were nearly the color of common
brown paper throughout, legs, wings, and body
being of much the same hue. They arrive every
year at the same time and in about the same numbers. They last a week or so, and although we
found them the entire length of our subsequent trip,
their favorite locality seemed to be the Sault. They
are used as bait for the lake herring, which I believe
is identical with the cisco, an excellent fish closely
resembling, and in my opinion equal, if not superior
to the white fish.

The trout usually begin taking the artificial fly in the early part of July, but although we had been warned that they were not as yet rising this year, we had no anticipation of the wretched luck that awaited us. Notwithstanding the water seemed promising, and deep, dark holes, beautiful eddies, and lively pools indicated success; and notwithstanding continual changes of our flies, we only killed three small fish. Perhaps the numerous natural insects, or the *larvæ* from which they were metamorphosed, proved a sufficient and preferable food; we could not induce the trout to rise, and did not even see them breaking.

Exploring all the little streams of the Canadian side, hoping at every cast to improve our luck, we worked our way slowly and arduously, for the water was unusually low, against the current, and steadily ascending with the strenuous efforts of our canoe men, who used stout poles for the purpose, we at last emerged above the islands and at the head of the rapids.

Here the water of the lake, confined to the narrow channel, chafed uneasily in tiny wavelets, as though conscious of the approaching struggle. Above, the river stretched away to the westward, evidently from a considerable elevation but comparatively smooth; nearer, it was rushing like a mill-race; below it was broken into white waves, huge cascades, and seething rapids. How wonderful is the change in the appearance of water lying calmly in the lake, hurrying rapidly but silently down a smooth slope, lashed into billows by the wind, toiling among rocks or leaping over falls—but above all is it peculiar and terrible in passing through broken descents! See it glide so deceitfully smooth, but with such resistless power toward the rapids; notice its tiny innocent ripples and childlike murmurs at your feet; see the pretty rolling undulations. Trust yourself to its seductions. Now it has you in its fearful current, now it drags you along, it clasps you struggling and shrieking in its fierce embrace; it throws its white arms around you, lashes itself into a fury, whirls you about in its powerful eddies, sinks you down in its mighty whirlpools, dashes you against the rocks, drags you along the jagged bottom, tosses you over the cascades, and finally flings you torn, bleeding, disfigured, and lifeless to the bottom of the tranquil pool at its base.

In the sunlight it resembles liquid crystal; flowing along placidly, transparent as the diamond, it sweeps upon the rocky shoals and flies up in a shower of purest pearls, alternately revealing or hiding some monstrous gem to which it lends its reflective brilliancy; over the limestone it is opal, over yellow rocks it becomes onyx, over the red, ruby or garnet, over the green, emerald.

Bending and waving in ever varying beauty of form, but carrying in its bosom or reflecting from its foam the sunlight fire, a thousand times intensified, of precious stones.

As the day was well advanced, we determined to trust ourselves to the unreliable element and run the rapids, which is one of the favorite amusements of the adventurous. This can be made as dangerous as desirable, according to the selection of route, either near shore, where there is only the chance of an upset and a few bruises, or through the centre, where it is certain death. We chose a middle course, but as near the centre as our guides, who were not venturesome, would go. Crossing over above the broken water to the American shore, the large, high-sided, but fragile canoe was headed down stream, giving us a view of the prospect before us.

Great ridges of white foam stretched at intervals almost from shore to shore, while the darker water was broken into heavy waves, curling up stream and ready to pour into the boat as it should rush downwards through them. At first the canoe settled gently, making us plainly feel that we were going down hill; then it gathered way as the current increased, and went plunging on its course. The waves flew from our bow or leaped over in upon us, the rocks glided by racing up stream, whirlpools twisted us from side to side; we sprang over tiny cascades or darted down slopes deep and dark, or shallow and feathery white with foam; we rushed upon rocks where inevitable destruction seemed awaiting us, and the shore, trees, and houses went tearing by; past the little island at the head of the rapids, past the main fall, through foam and spray, we dashed headlong, till the few minutes required for the entire descent being exhausted, we glided calmly and quietly into the water below.

Looking back it seemed as though we gazed upon a hill covered with water instead of up a river, and nothing but practical experience would convince a tyro that it could be navigated in safety with a birch canoe. Exhilarated with the pleasurable sensations we had enjoyed, and satisfied that the trout were not in a rising mood that day at least, we returned to the hotel.

The few fish we had killed were transferred by our host to the cook, and reäppeared on table in fine style. After discussing an excellent dinner and comparing notes with the other fishermen present, we accepted the invitation of the canal superintendent to examine the locks and visit his pond of tame trout. We found the canal an admirable structure, expensively built, and of a size to accommodate the largest steamers that navigate Lake Superior; not, however, being skilful in works of that character, we felt more interest in the trout pond.

The latter was quite small, fed by a pipe from the canal that cast up a jet in the centre, and was filled with over a hundred of fine, large, active trout, weighing from one to four pounds. They were wonderfully gentle, would feed from the hand, allow any one to scratch their sides and lift them from the water, and if one end of the food was held fast, they would tug like good fellows at the other. When we held a piece of bait between the first finger and thumb, and at the same time presented the little finger, they would frequently seize the latter by mistake; and although on that occasion they let go instantly without doing the least harm, the proprietor said when hungry they occasionally left the marks of their teeth. It was extremely interesting to watch their movements, as their appetites were never allowed to become ravenous and produce quarrelling among themselves. They were magnificent fellows, swimming about majestically, and coming to the surface in a fearless way to return the gaze of the spectators.

The trout were mostly taken in nets from the canal when the water was drawn off. They had been known to spawn, trying to ascend the jet for that purpose, and depositing their eggs where the water fell; but the spawn either was eaten by their comrades or failed to hatch. Under no circumstances, however, would the young have lived among such rapacious giants.

Having amused ourselves sufficiently with the tame trout, we turned our attention once more to their wilder brethren; but as no better success attended us than in the morning, we returned early to superintend the capture of the white-fish. Every morning and evening the Indians and half-breeds are seen by pairs in their canoes, one wielding a large net with a long wooden handle, and the other plying the paddle. Ascending cautiously to the eddy below some prominent rock, the net-man in the bow peers into the troubled water, and having caught sight of the white-fish lying securely in his haven of rest, casts the net over him. The moment the net touches the water the other ceases paddling, and allows the canoe to settle back with the current; the fish thus entangled in the meshes is lifted out and thrown into the boat. The net is about four feet across, the rim is of wood, and the handle is bent at the end so as to afford a secure hold. Nothing but the practised eye of the native can distinguish, amid the foam and spray and broken water, the dim and varying outline of the fish. Many are frequently taken at one cast, and they are sold, large and small, for five cents apiece.

Although undoubtedly delicious eating, fresh from the cold water of Lake Superior, white-fish are not superior in flavor to their smaller brethren, the lake herring. The latter, so closely resembling the former as to be only distinguishable by the sharper projection of the lower jaw, are taken with the natural brown fly that has been already described. Differing little, if at all, from the cisco of Lake Ontario, they rise with a bolder leap at the natural fly, and their break is as vigorous and determined as that of the trout. They do not seem, except on rare occasions, to take the artificial fly, but with bait not only furnish pleasant sport for ladies, but an admirable dish for the table.

The lake herring is found in many of the extensive waters of the West, but being smaller than the white-fish, is overshadowed by the reputation of the latter. It is a pretty fish, bites freely and plays well, but having to contend in delicacy against the white-fish, and in vigor against the trout, it does not receive the attention it deserves. Early in July.

they collect at the Sault in millions, filling every
eddy of the rapids and crowding the canal, and devour the dead and living *phryganidæ*. Later they
retire to deep water.

It being now apparent that the trout did not intend to accept our delusions as veritable insects,
and as fish of three and four pounds had been taken
with minnow, much to our envy, Don determined
to try the bait. There are several species of minnow captured from among the rocks of the Sault
in shrimp-nets, but the favorite is a peculiarly shaped
fish bearing the euphonious title of *cock-à-doosh*.
What the name signifies, either in French or Chippewa, we could not ascertain ; but the broad, round
head and slim tail remind one of a pollywog, which
of all created things it most resembles. The cock-
à-doosh is a muscular little fellow, and not appearing to mind a hook thrust through him, furnishes a
lively, attractive bait.

At the suggestion of some gentlemen who were
old habitués, and who recommended to us a couple
of men that had accompanied them on former trips
up the lake, we had determined to discard our present boatmen, although without cause of complaint,
and engage Frank and Charley Biron to accompany
us into the woods. We had laid in our supplies of
food, all of which, except the tent, the liquor, solidified milk, and a few especial luxuries were purchased
in the village stores, had made our preparation for
departure in the morning, and devoted the afternoon
to fishing the little rapids.

Our present men had already ascertained our intended change, and we had hardly pushed off before old Joe began upon us. He spoke French, the language of communication between the natives and travellers, and never shall I forget his reproachful tone and manner. Perfectly respectful, he pictured our enormities and unkindness in such eloquent words that we hung our heads in shame.

Never before had he, the chief of the Melicetes, acquainted as he was with the whole length of the lake, been displaced for younger men. The young men were good voyageurs—that he did not dispute; but was it reasonable to prefer them to one who had lived his whole life in the woods, or was it right to brand with disgrace a guide who for two days had served us, as we admitted, faithfully? Unusual, indeed, was it to change the men, and should he have this discredit cast upon him? He had not been engaged positively to accompany us; but had we not spoken to him and asked his advice? Was he not justified in expecting it? He was sorry and hurt that we should have done so; he had been pleased with us; he knew that he could have pleased us; but could he rest under such an imputation? Were younger men better boatmen than he? Were they better acquainted with the lake? Were we dissatisfied with him so far? Why, then, had we changed, unless indeed to offend him? His feelings were wounded, and he felt sure that we must regret our injustice. If we said that we had been advised to do so, it must have been by persons who

3

did not know him or had some unworthy object; and should we have done so great a wrong without more inquiry? "No, *messieurs ;* this is the first time I have been turned away for younger men."

It is impossible to give his language, for Joe, although usually taciturn, burst forth with an overwhelming flow of eloquence, showed us our conduct in such a light that we would gladly have retracted, and compelled us to take refuge behind our ignorance of the customs of the place. Disclaiming the intention to cast a slur upon him, we expressed the fullest confidence in his abilities, and said that were it not too late we should cancel our other engagement. Somewhat mollified, the pleasant expression returned to the old brave's countenance ere we reached the little rapids, where the excitement of fishing diverted our attention.

Don here met with his first success with the cock-à-doosh, striking and killing, after a protracted struggle of twenty minutes, a fine trout of three pounds. The rapidity of the current, which flowed deep and strong without an eddy, gave the fish a great advantage, and tried the rod to the utmost. The hook, from its size taking a better hold than the diminutive fly-hooks, remained firm and enabled Don at last to bring his prey to the net—and kill our first large fish in the waters of Lake Superior.

Having fished faithfully, but in vain, for a mate, although we saw in a deep pool quite a number as large or larger, and as my fly would still only attract the small ones, we headed once more up-stream.

The two miles' return was slower than our descent, and gave us time to admire the scenery, to watch the vessels passing through the narrow channel of the shallow river, and note the decaying woodwork of the old fort that once did good service against the Indian, but would be a ludicrous structure in modern warfare. On arriving at the Sault the finishing touches were given to our preparations for camping out, and a wagon engaged to transport our stores by land to the head of the canal, where our new men and their barge were to meet us early on the morrow. We parted with Joe, who, however, that evening and next morning heaped coals of fire on our heads by doing us innumerable little favors in the way of suggestions, advice, and physical aid.

The day following, as the last article was placed upon the cart, we were informed that neither eggs nor bread was to be had in the village. Our horror, or rather mine—for Don little knew what a dearth of eggs implied—can only be appreciated by an experienced cook ; bread was a minor matter, as we had ship-biscuit, but eggs were indispensable. It appeared on inquiry that the baker had been heating his own coppers, as the fast men express it, instead of his oven, and was now sleeping off the effects of his debauch ; and hens, feeling their importance in that desolate country, only lay on special occasions.

While we were in a condition bordering upon despair, uncertain whether to proceed, the steamer *Illinois* hove in sight. Never was an arrival more opportune, for one of the numerous ventures of the

bar-keepers on these vessels is to supply the country
with eggs, and recollections of the baskets full that
we had seen hanging from the cross-beams of the
City of Cleveland came vividly to our minds. Leaving Don to purchase the eggs, I pushed on with the
baggage. The former boarded the steamer as soon
as she touched the dock, and, rushing to the bar-
keeper, demanded eight dozen eggs. He was informed, however, that they were sold by the basket,
which contained fifteen dozen, and he could have no
less. Then it was that Don rose to the importance
of the occasion. Others might have doubted, hesi-
tated, or failed to make the purchase at all; but he,
without a pause, grasped the basket, laid down the
money, and started for the head of the canal. Fifteen dozen eggs were a perfect mine of comfort;
in their golden bosoms lay undeveloped numberless
egg-noggs, delicious cakes, and appetizing omelets,
and Don's character was established for ever.

The wind, strong and contrary, was dashing foam-
crested waves against the piers of the canal, threatening to make our journey a slow one; our goods
and chattels were safely and carefully stowed, filling the barge as nearly as was desirable; we had
even cast off and commenced our voyage, when
through the canal we saw approaching a tug-boat.
She was called the *Bacchus*, and, like her jolly pro-
totype, willingly lent us her aid; and giving us a
tow, made our old boat, for that occasion at least, a
fast one. She tore her way along, crushing the
waves with her high bow, throwing a mass of white

water from her propeller, and carrying us in fine style past *Pointe aux Pins*, nearly ten miles of our route.

Having left her, as our course now lay more to the northward, we managed with hard rowing, very different from our previous gallant progress, to reach *Pointe aux Chênes* or Oak Point, in time for dinner. Looming up at the distance of about six miles, rose abruptly to the height of five hundred feet the bold promontory of *Gros Cap*, its round head enveloped in driving fog. A scanty verdure of pines and firs covering its sides, it stood out a bold landmark, being the first high land of the northern shore.

About half-way between Pointe aux Chênes and Gros Cap lies a low and narrow island, covered with small trees and underbrush, furnishing an admirable camping-ground; and the wind increasing as the fog descended, crawling slowly down the mountain sides, we could advance no further.

All day long canoes filled with Indians, taking advantage of the to them favorable wind, passed us on their way to a grand council at Mitane. It was wonderful where they could all come from; the men seemed to carry their wives, papooses, and household gods, and were accompanied by numberless dogs that ran along the shore; one party consisted of a squaw seated at the bow to paddle, another in the stern to steer, and a brave amidships fast asleep; the canoe was propelled by a blanket, used as a sail. The Indians exhibit great skill in

sailing so unsteady a boat as a canoe ; although to
ordinary mortals it is difficult to stand up in one,
they manage to sail them in heavy winds and over
a rough sea. This art appears to be peculiar to
them, for I have never known it attempted by the
Canadian voyageurs, nor even by the half-breeds.

The fogs rising from the cold waters of Lake Superior are frequent and dense ; on this occasion the
moisture settled upon the bushes, fell from the
leaves in large drops, and dampened the boughs of
which our bed was to be composed. For this latter
purpose, as there was no sapin on the island, we
were compelled to use oak sprouts, a substitute
that Don at first, attracted by its beauty and apparent comfort, approved, but which, when before
morning the leaves were pressed flat and the stems
made unpleasantly prominent, he anathematized
vigorously.

After supper we wandered along the shore, pick-
ing up the queerly shaped and oddly colored stones
that abound on the Canadian side of the lake. No
agates nor amethysts, and none of the really beautiful pebbles, are to be obtained south of Michipicotten, but everywhere are curious specimens to be
found. Carried, as it is supposed, by the ice-drift
of former ages from their natural beds, crushed by
the moving mass, and rounded by the beating waves,
the hardest only survive, while the strangest and
most incongruous varieties are collected together.
Meeting with novel specimens at every step, we
were continually rejecting what we had just selected,

till we hardly knew which were really the most remarkable.

Next morning broke with the weather the same,
but towards mid-day the wind fell. Don had been
gratified with his meals thus far, but on being offered rice for breakfast, said that it reminded him of
his European experience, where rice was not considered fit to eat without being filled with raisins
and having goose-gravy for sauce. In fact, he did
not think he could eat it without these accompani-
ments. Before the trip was over, however, he found
that in spite of European authority and the absence
of goose-gravy, rice was quite palatable.

By hard work we reached the camping-ground at
Gros Cap, a small island almost adjoining the main
land, which is too rocky and precipitous to locate a
tent, and having arranged our camp amid the driving fog, essayed the fishing off the point. Fortune
did not smile upon us; and having killed one fish
for supper, we were glad to escape from the cold,
damp air, and return to the warmth of the fire.

The appearance of the rocks in this region is remarkable. Not only are they veined with metal
and quartz, running in long seams, but they are cut
up by deep furrows, at the bottom of which are
strewn broken and pounded stones. The origin of
the furrows, or scratches as the geologists term
them, has been differently explained; some writers
attributing them to the action of water, and others,
with probably the correct theory, alleging they were
made by the ice-drift of former ages. The ice-drift

was the accumulation of snow and ice in the neighborhood of the north pole, its increasing masses forcing their way towards warmer latitudes, and carrying with them immense rocks and boulders. The drift formerly extended far beyond its present limits, pouring into the deep water of Lake Superior, and must have crushed and riven whatever lay in its course—cutting deep furrows whenever the boulders it was carrying came in contact with the unyielding native rock. The character of the rifts, which do not resemble the effects of water, their uniformity of direction, and the pounded character of the stones, confirm this view.

Whatever may have been their origin, they are troublesome to cross, forming as they do abrupt gullies running from high up the hills into the deep water, and occurring at every few hundred feet. But where they pass below the surface, they and the natural caverns worn by the waves form admirable retreats for the timid trout. For the whole length of the shore, the broken rocks lie piled up in the water, and at some places extend far out; as they furnish the best locality for sport, although generally the angler has but a short distance to cast, occasionally a long stretch has to be made. The wind is frequently adverse or across his line, and as he must reach a particular spot in spite of all obstacles, his capabilities are often put to the severest test.

To encounter and overcome difficulty is the true sportsman's delight, almost as much so as to see the silver-sided beauties of the lake rise suddenly from their fairy caverns and seize his fly, to feel them struggling and fighting for their liberty, jumping again and again, and finally to watch their fading brilliancy enveloped in the fatal net. The trout of this region resemble the sea-trout of the Gulf of St. Lawrence in their habits and appearance. They have the same pearly whiteness on their sides and bellies, heightened by the minute specks of carmine; the same vigor and dauntless courage, the same savage voracity, and the same way of springing out of water when they are on the line. They rise unexpectedly with a rapidity resembling fury, grasp their object with determination, and on being struck, fight bravely. Their flesh, also, is equally red and firm, their fins of a pure color but not quite so delicate, and their shape identically similar. Of course they could never have ascended from the sea, but are indebted for these peculiarities to the pureness of the water of the lake, as the sea-trout are to that of the gulf. And whereas the sea-trout lose their brilliancy on ascending the rivers, so do these of the lake—a fact which we afterwards ascertained—becoming even darker colored than their brethren of the lower regions, and obtaining the reputation among the ignorant natives, from their changed appearance, of being poisonous.

Another party of fishermen had located on Gros Cap island, our tents being pitched within a few yards of each other, and we passed a pleasant evening in their society; our pipes—for I had after much difficulty persuaded Don that cigars were

made for the club-house, not the wilderness—suggested inquiries about the native weed called Kinnikin-
nick, which the Indians in their grand peace councils
used before the advent of the white man, and which
in a perverted form had lent its name to the tobacco
we were using. It appeared that the identical weed
was growing close around us, and although the Indians of their party laughed with contempt at any
one using it when pure tobacco was to be had, we
induced them to collect and prepare a small quantity.

The preparation consists of drying it thoroughly
by the fire until it is brown, and then pulverizing it
by friction in a cloth. The operation was soon
completed, but, although we tried it mixed and un-
adulterated both, we were forced to admit it had
absolutely no flavor whatever. Perhaps it wanted
more time or care in the curing, as the men complained of the dampness.

Our new-made acquaintances left next morning
early, and Don and myself took a late breakfast and
were joined by an unexpected visitor. A quantity
of cold potatoes and ship-biscuit, intended for our
men's breakfast, had been temporarily placed on a
neighboring log, and while we were partaking of
warmer edibles, a few steps off a pretty little ground
squirrel ran out, chirruped a merry good-morning,
and proceeded as a matter of right to help himself to
the cold victuals. He was sleek, bright-colored, and
fat, evidently accustomed to many such repasts;
and after trying a piece of potato and finding it

was good, he took up a whole one in his mouth and
ran off with it. It was larger than his head, and
looked droll enough in his mouth, stretched to the
utmost; he had not gone far before his sharp teeth
cut through, and taking out a piece, let the rest fall.
Not taking the trouble to pick it up, he returned
with another little cry to the dish, and this time
chancing on a smaller one, carried it off in safety.

Having stowed that away, he returned, and being
satiated with potatoes, tasted the biscuit, which had
been soaked in grease and was tender. The piece
he selected had a larger piece hanging to it, and to
see him pull the latter off with his fore-paws was
highly amusing. The biscuit, on trial, proving acceptable, with a little flirt and another cry, he
seized quite a large piece, and with a glance at us
as much as to say, "I am only taking a fair rent for
the use of my land," he ran off with it in the same
lively, confident way. It was a beautiful sight, and
we stopped our meal to watch his pranks.

MOUTH OF THE AGAWA.

· CHAPTER III.

Gros Cap is the first of the rocky hills that form the northern boundary of Lake Superior, and which, with the higher chain of mountains further inland, divide the streams that run to the southward from those that empty into Hudson's Bay.   The Hudson's Bay Company, that wonderful commercial undertaking that had stretched its arms across our continent, and which, after the destruction of the beaver, has lost its influence and been shorn of its power, has stations along the coast of Lake Superior at the mouths of the various rivers of importance.   At the Sault on the Michipicotten, the Pic, and the Neepigon, they have planted their trading posts, and although their glory has departed, they are still kept up and do some business.   These stations were convenient stopping-places for the voyageurs, and were located at the mouths of rivers, of which the fountain-heads communicated by a portage with a different system of waters.   For instance, the Michipicotten is the Indian highway to Hudson's Bay, and both on it and on the rivers adjoining that empty into the latter, has the great Company its stations. The study of the results that that purely commercia undertaking has achieved, from the Saguenay River throughout the British Provinces to the far West, is an instructive evidence of the power of man unrestricted and untrammelled. In various ways it has left its mark for ages.

Gros Cap is a perpendicular bluff, shooting straight up from the water, and with its rocky clefts just furnishing foothold for the active fisherman ; pieces of rock seem to have been broken off and thrown into the water at its base, and among these trout are numerous. No place furnishes a pleasanter camping-ground, although not directly at the fishing ground, and few spots afford better sport. As fortune was not particularly propitious, and our journey was indefinitely extensive, we took advantage of a calm that had settled down upon the lake to push on across Goulais Bay, which lay as calm as a mirror, bathed in the glorious reflection of a cloudless sky.

Farther out, Isle Parisienne seemed floating on the water, while inside of us the bleak sides of the abrupt hills were reflected in long wavy lines. The sun had climbed the eastern sky and poured down a flood of warmth and light in strange contrast with the tempestuous weather of several days. The atmosphere, instead of being dense with impenetrable fog, was exquisitely transparent, and the water, that perfect ornament to every landscape, stretched away as far as the eye could reach.

> " Dark behind it rose the forest,
>   Rose the black and gloomy pine-trees ;

Rose the firs with cones upon them,
Bright before it beat the weather,
Beat the clear and sunny water,
Beat the shining Big-Sea-Water."

Such a day is admirably adapted for taking lake-trout, and no sooner had we entered the bay than our lines were arranged for the purpose.

The Namaycush—pronounced more nearly like Namægoose, with the accent on the second syllable—the *Salmo Amethystus* of our ichthyologists, the *Truite du Lac* of the Canadian, and the *Mackinaw Salmon* of the American, inhabits Lake Superior throughout its length and breadth, is captured along the shores and in the bays, and when smoked, furnishes the principal food of the Indian. It prefers a rocky uneven bottom, where the water is neither excessively deep nor very shallow, and during the summer months bites readily at any of the ordinary trolling-spoons. An ivory imitation-fish is especially attractive; and an old-fashioned bowl-spoon, elongated with bright tin on one side and red on the other, is in general use.

Whenever the Indian is paddling in his canoe over any of the favorite localities, he trolls with the latter bait, which is sold at the stores in the Sault; and to make it imitate more accurately the herring it is intended to represent, he attaches the line to his paddle. By this means a peculiar darting motion is given to the spoon which is said to be very fatal. Buel's patent spoons, whether with feathers or without, are successful; and so little particular is this voracious fish, that he will bite at a white rag attached to the bare hook.

Once struck, however, and he surrenders without an effort, appearing even to swim gently forward, which conduct, although natural in a man under similar circumstances, is not expected in a fish. So slight is his resistance that it is difficult at times to tell whether he is on the line or not; and although, of course, on approaching close to the boat he flounces and struggles a little before he can be gaffed, he affords the sportsman no excitement whatever. He may also be taken in deep water with a long line and sinker, with the lake-herring for bait, and is thus during the fall captured of enormous size.

He is found occasionally to weigh seventy pounds, and perhaps more; a handsome fish to look at, he is also excellent to eat, and with the peculiar conformation of the trout, he combines its elegance and the rich redness of flesh of the true salmon. He is rarely taken by trolling to exceed ten pounds in weight, and on the north shore more frequently of five or six; but of that size is an invaluable addition to the fisherman's larder. He may be either boiled or broiled, and makes a capital foundation for a chowder. He must by no means be confounded with the siskawit, which is only taken in the upper part of the lake, rarely exceeds seven pounds, and is so fat as almost to dissolve in the frying-pan—at least we were thus informed by our guides, for we took none ourselves.

The best time to take them is in calm weather,

because on such days they rise nearer the surface and are able to see the bait farther. If the wind is strong or the boat moving rapidly, they will not bite; in fact, the boat should not be sailed or rowed faster than three miles an hour, and a common hand-line of fifty or a hundred yards is sufficiently good tackle. They are persecuted by the aborigines, who capture vast numbers for winter use; but we never caught more than a dozen in a day, as we never fished exclusively for them.

Goulais Bay is one of their favorite haunts, and we were soon made aware of their presence. I had the pleasure of striking the first, and felt some anxiety, it being a new species to us, till he was safely gaffed and landed. He weighed four pounds and a half, and we fairly feasted our eyes over his beautiful shape. Don soon had one still larger, and we took six while crossing from the headland of Gros Cap to Goulais Point. They differed a little in size, the largest being six pounds, but not in shape or appearance, and were in their way as exquisite a collection of fish as ever were taken.

We could doubtless have killed many more if we had wished to remain for the purpose; but the Harmony River, our destination, was a long way off, and the sun was running across the sky at a rapid rate.

We stopped to dine at Goulais Point, and took advantage of the opportunity to bathe; the water, close to the shore where it was shallow and had been heated by the sun's rays, was warm, but occasionally streaks cold enough almost to freeze the blood were encountered. The Namægoose, on being prepared for the pot, were found to contain spawn well advanced, and were exceedingly fat.

The dinner being over and the men rested, our slow progress was resumed, and we passed Maple Island—*Isle aux Arabes*—into Batchawaung Bay. The sun in his downward course marked out a broad golden path upon the still surface of the lake, vividly recalling to our minds that most exquisite picture in "Hiawatha" of the chieftain's departure for the "land of the Hereafter;" which now had the charm of a peculiar interest, as we were floating upon the very waters where the scene is laid:

> "And the evening sun, descending,
> Set the clouds on fire with redness;
> Burned the broad sky, like a prairie,
> Left upon the level water
> One long track and trail of splendor,
> Down whose stream, as down a river,
> Westward, westward Hiawatha
> Sailed into the fiery sunset,
> Sailed into the purple vapors,
> Sailed into the dusk of evening."

Thus dreamily murmured Don, as with his back against our biscuit-barrel, and his feet upon our butter-tub, he gazed upon the dying glories of the orb of day; and now, as the last glimmering spark sank below the horizon, the strange pale light of the north crept over the sky; the stillness of death brooded upon land and water, and *ephemeræ*, issuing from their *larva* state, burst into winged life and

followed the course of our boat. Fronting us was
the long island called by the same name as the bay
beyond it, and towering far above were the mountains of the mainland, cleft in two places where the
Harmony and Batchawaung Rivers had broken their
way to the lake; to the right extended the bay for
many miles, and to the left stretched in its immen-
sity the trackless "Gitche-Gumee, Big-Sea-Water."
Darkness approaches slowly in northern latitudes;
our oarsmen were weary, and our pace was moderate, but we had to make a long detour to reach
the river beyond, and it was determined to camp on
the island. Reaching the upper end, we landed, and
our men searched for a favorable spot. One peculiarity of a voyageur is his antipathy to camping at
an unusual place; warned by his experience of the
inconveniences that attend such a course, the difficulty of making a comfortable bed, properly securing the tent, and arranging the fire, he will endure
considerable extra labor to reach a spot with which
he is acquainted. Therefore we were not surprised
when Frank reäppeared and announced the imprac-
ticability of establishing our camp.

The day had been hard for the men; the weather
had been hot and the journey long, and it gave me
pleasure to hear Don propose that we should row
for a time. He was rather unaccustomed to the exercise, but kept up bravely as we continued our
course round the island and across towards the main
shore. The pale light still filled the atmosphere to
that degree that, at nine o'clock, we could read fine

print; the *ephemeræ* still followed us with fluttering
wings, and whisks extended; the death-like calmness
still rested on the unruffled water. At the point of
the island were four pretty little islets clustered together, lending additional beauty to the bay embosomed in majestic hills. The way seemed lengthened out amazingly, and our arms were weary, and
the night had closed in darkness ere we reached the
mouth of the Harmony River, the *Auchipoisæbie*
of the Indians. Here we found an old camping-
ground, almost a cleared field in size, and the remnants of several wigwams. Collecting the poles of
the latter, we built a rousing fire that illuminated
the surrounding forest and cast a lurid glow upon
our active men. By its light we landed our stores,
pitched our tent, established our quarters, and retired to rest.

We had made a long thirty-five miles, against
unfavorable circumstances, felt exhausted but
thankful we had arrived at last, and taking a little
refreshment, drank good-luck to ourselves and the
Harmony. Just as I was about closing my eyes to
sublunary things, Don remarked:

"There is a serious question I have to put to you.
To-day's journey has probably been exceptionally
slow and tedious, but how long, under ordinary circumstances, do you think it would require to come
from New York to the Harmony River?"

Next morning early having broiled a Namægoose
for breakfast and found it both well cooked and excellent, we ascended the level water that extends

for some distance from the mouth of the river. The day was fair and the wind favorable, the birds sang their welcome merrily, and the trees bowed gracefully as we passed. An old duck and her young were startled by our approach, and fled, making such use of their powerful legs as to outstrip us readily. A short distance beyond the smooth water, and almost three miles from the lake, we came to the lower fall or pitch of the stream, which had become quite narrow, and there we made our camp.

It was a lovely spot; the thick trees formed a dense shade over our tent, the trembling cascade furnished continual music; opposite, a rivulet of purest ice-water emptied into the stream; in front the river spread out into a broad, quiet pool; while through intervening trees and bushes we could catch glimpses of the high falls a few hundred yards above us. Previous camps had been located at the same place, and a path had been cut to the rock close by, from which we could fish below the cascade.

Hastily disembarking such things as we had brought with us, impatient to explore the river, and tantalized by half glimpses of the cataract beyond, we crossed the stream in the barge, and guided by Frank, followed a well-worn pathway in the woods. A few hundred steps brought us to the bank, where a glorious prospect greeted us. The stream, rising among the summits of the hills, pitched down over a slanting precipice, seaming its brown face with irregular, delicate lines of silver. Issuing from a mountain gorge, so far above as to be scarcely dis-

tinguishable, it leaped over pitch after pitch, collecting in deep pools at every break, and whirling round
or dashing over huge boulders in its course, till descending the last shute, the main body tumbled in
one heavy wave into a dark, turbid pool at the base.
From either shore the evergreen trees projected, lean-
ing over as if to protect the uneasy river, and a
heavy trunk, originally torn up and borne along by a
spring freshet, had lodged upon a broad, bare, rocky
island in the centre. Numerous little rills branched
off from the main stream, and forming innumerable
fantastic miniature water-falls, sought different paths
to the lower level. The rocks were bare and mostly
of a dull brown, constituting a strong contrast to
the green fringing of the mountain sides, and were
worn away by the immense volumes of water and
ice that forced their way through in early spring
and swept them clear of vegetation.

At the foot of the lower shute there was a seething cauldron, white with foam near the fall, and
black from its great depth in the centre; below, the
wearied stream rushed down a stretch of rapids, and
sought temporary relief in a broad, quiet basin that
reached to the first of the cascades, close to our
camp, and in which the water seemed absolutely
motionless.

Hardly giving ourselves time to note and enjoy
the beauties of this most romantic spot, and urged
on by the sportsman's instinct that looks to the attractions of nature, after having tried for game, we
commenced casting in the rapids. Our efforts were

rewarded, and we landed some fine fish of from one to two pounds, and had grand sport with them in the current and eddies. Putting on for a tail-fly a large, full, brown hackle with scarlet body and silver twist, I at last advanced cautiously towards the black pool below the shute, and keeping well out of sight, cast it across the boiling water; it fell among a mass of whirling foam, but being swept down, passed over a portion of the dark water, and was ravenously seized by a fine trout.

Astounded at the unexpected consequence, the frightened fish darted hither and thither about the pool until, finding his efforts to free himself vain, he rushed towards the rapids below. Here the rod and line were powerless to restrain him, and he made the reel spin as I followed along the rocks. However, with care he was guided through the dangers of the foaming current, strong eddies, and projecting rocks, and was led after a long battle into a spot of comparative quiet, near an old dead tree that projected over the water.

Being myself prevented from approaching by the branches of this tree, I instructed Frank to watch a good chance and use the net; but never shall I forget his look as, after two or three vain attempts —for he was not altogether skilful—the upper fly caught in his shirt, and the trout, which must have weighed at least three pounds, made a furious dash, parted the leader, and escaped. As though it was my fault, instead of his awkwardness, Frank turned towards me with a most reproachful expression, and without a word came to have the hook cut from his shirt, intimating that if I would hook him, I could not expect to land large trout.

The fishing below the falls of the Harmony was absolute perfection; although the fish were not large, that is, not of monstrous size, and rarely exceeded two pounds, they invariably after a short struggle took to the rapids, and compelled us to follow them, at a pace and under difficulties that brought salmon-fishing vividly to our recollection. The steady roar of the falls and the picturesque wildness of the scene added to the intensity of the enjoyment, and served to occupy our minds when not employed upon our sport. Of easy access from our camp, we afterwards ordinarily visited them alone, leaving the men to attend to numerous household duties, and had the advantage of being able to wait upon ourselves.

The hours passed quickly by, and when the calls of appetite could no longer be resisted, we found ourselves with two dozen splendid trout, which were the selection from nearly a hundred. Well satisfied, we hastened back to our camping-ground which Charley had been busily arranging, and while the men were preparing dinner, we tried the cascade near by.

This was certainly a fortunate day, for Pedro soon hooked a splendid black bass and landed him, after a vigorous struggle of half an hour; he weighed three pounds and three-quarters, and was thoroughly game, and established a fact that Professor Agassiz seems to doubt—that black bass inhabit Lake Superior. The guides recognised him at once as an old acquaintance, and called him by the familiar name of *achigon*.

After a hearty dinner we descended to the mouth of the river for the residue of our camping articles, and while returning I trolled with a small Buel's spoon. Unfortunately happening to espy a duck upon the water, I laid down my rod to take the gun, when a black bass struck, nearly jerking the rod out of the boat, and with a mad spring carried off my bait and casting line, while the duck, alarmed at the noise, flew away amid the confusion.

Having landed our load, and leaving the men to complete the camp, Don and myself hastened back to the scene of our morning's sport to renew, and even surpass, our previous enjoyment; for after killing several fine fish in the strong water in splendid style, I struck one of great weight in my favorite pool. He soon took to the rapids, and stopping in an eddy, fouled the line without escaping. In vain all means were tried to clear the line without alarming the fish; it had caught on the further side of a large stone, and could only be reached from a rock that projected its smooth, slippery surface above the current at some distance from the shore. Rendered desperate, and summoning all my courage, I crept out into the rushing stream, and, supported by the handle to the landing-net, succeeded in reaching this dangerous location.

No sooner was the line free than the fish again

darted down stream, taking out the line at a tremendous rate. I turned to follow, but what was my dismay to find that, although I had managed to get from the shore to the rock, the current followed such a direction that I could not return. On went the fish; in vain I sounded the bottom with the handle of the landing-net, or felt for a safe footing, or essayed to jump; the water was too threatening and the risk too great. Still the fish kept on, and I had just made up my mind to take the leap for his life or my own, when the line became exhausted and the leader parted. Slowly I wound in the line, sadly picturing the supposable weight of the escaped fish, and depressed in spirit, managed with Don's assistance to regain *terra firma*. The only consolation was in the thought that we had secured full as many fish as we could use.

That night was extremely warm, and one of the most trying I ever endured in the northern woods; not only were mosquitoes abundant and ferocious, but that terrible pest, the sand-fly, existing by myriads in the sandy soil, made merciless attacks upon us. The shores of Lake Superior are unpleasantly prolific in all the minute torments that are most dreaded by the sportsman. During the day the black-fly absolutely swarms, in the evening the sand-fly arises from the sand in invisible millions, and at night numberless mosquitoes continue the pursuit; repelled, but not dismayed by ointment and liniment, they wait till it is dried or rubbed off, and dart upon the exposed part; they far exceed in

4

numbers their brethren of New Brunswick, where the rocky soil is less suited to them, and, in spite of all defences during hot weather, inflict much misery.

Don's first idea was to despise their attacks, and, disbelieving the virtues of pennyroyal and creasote, stoically to endure the discomfort of the woods as a necessary accompaniment to enjoying the pleasure; but by the time tea was over he had changed his mind, and at bedtime carefully enveloped himself in his veil.

The thermometer rose to eighty-six in the tent, and being little lower at midnight, the veils were found to be rather suffocating. The moderate temperature of the northern climate is the great protection of the sportsman; ordinarily in a trip of a month there will not be three oppressive days, but when the weather is warm and insects numerous, a good chance is offered to exhibit courage and jollity. Next morning, when the heat continued, and the sun, rising above the hills, shone through the dense fog like a globe of fire, Don wore a solemn but patient expression of countenance, and fully justified my confidence in his endurance.

The weather during the early season had been warm and dry, and the lake was two feet below its ordinary level, and although its main body retained a cool temperature, the shallows were heated. The rivers, on the contrary, that flow into it from the north, taking their rise from swamps and shallow ponds, not only are tinctured with decaying vegetation and are of a rich amber hue, but had absorbed the heat, so that the fish which in our latitude are in summer accustomed to desert the lakes for the cool spring brooks, had mostly left the rivers for the cooler lake. Only where the water was cooled and aërated by a fall, or at the mouth of some trickling spring, were they to be found in any numbers.

I have said that opposite the camp there was such a rivulet, and at its mouth, crowded together, each striving to get his nose nearest to it, was a fine school of large fish. The water of this rivulet must have been not far above the freezing point in temperature, and was delicious drinking, while the main stream was nearly tepid.

Being informed by our guides that there was a second fall above the first, and good fishing near it, we proceeded, after taking a few fish and a good drink from our spring-water rill, to ascend the river. We were compelled to make our way through the brushes and undergrowth, over the dead trees, and among the rocks that covered the shore, and were hardly repaid for our labor; the fall proved to be only a small cascade, and though there was a deep fine pool at its base which Frank assured us contained trout of five pounds, we could not persuade any of them to rise. As no fish above the main fall could have access to the lake, I felt convinced there were none of large size, and the weather continuing warm, we returned early to the camp.

That evening was again devoted to the black bass, which took both the fly and spoon greedily, and which, when captured, were deposited alive in a pond-hole in the rock, where their appearance and motions could be studied to advantage. They were not handsome fish, with their broad backs, deep bodies, and thick heads; their extended fins were peculiar and characteristic, and their general form, fierce red eyes, and large mouths were more indicative of ferocity than grace. Those that we opened, although it was in the month of July, were heavy with spawn, and the ova had the appearance of being almost ready for deposit,—suggesting the possibility that these fish differ from those of the eastern country in their spawning season. It is hardly conceivable that they would carry their eggs till April or May of the ensuing year, in which month black bass spawn elsewhere; and if not, their habits must be entirely dissimilar.

The long walk through the sand and mud had made our shoes rather unpresentable, restoring along the edges the original russet of the leather; and as he was about retiring, Don suggested to me the propriety in our next trip of bringing with us blacking and brushes.

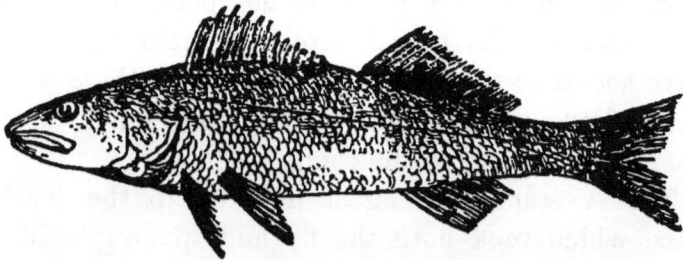

PIKE-PERCH.

&

CHAPTER IV

NEXT morning, the weather being cooler and the wind favorable, we took our departure, after having captured some fine fish at the falls pool, for the Batchawaung River. It was but a short journey round a sandspit that projected into the bay, where we took a single trout, and we were soon in the mouth of the deep dark river. The banks were low and of course covered with trees, most of which were of the deciduous character; the water was sluggish, and the interval between the bay and distant mountain extended several miles.

We passed an Indian paddling a canoe loaded with bark, the sole occupant besides ourselves of the quiet stream, and our guides conversed fluently with him in the musical Indian tongue. Occasionally a brood of ducks, alarmed at our approach, broke the oppressive silence with their vigorous efforts to escape, and Don, trolling with Buel's spoon for black bass, struck and landed a small ill-favored pickerel —*esox boreus*—of some four pounds weight.

The Batchawaung is the favorite resort for anglers who visit the north shore, and being within easy access of the Sault—not more than a day's sail with favorable weather—is fished to excess. It is a large stream, filled with rapids and pools, and usually

crowded with trout of immense size; but the water
is dark and easily heated, so that the fish often desert
it for the lake. There is a sameness about the
Batchawaung, and a want of picturesque effect, that
is altogether different from the Harmony; we missed
the noise of the falling water, the sight of the pretty
cascade, when we came to pitch our tent about four
miles from the mouth, at the first shallow rapids,
and throughout our whole trip we never saw the
equal of the romantic Harmony.

There are but two rivers emptying into Batchawaung Bay that are generally laid down on the
maps—the Batchawaung and the Chippewa—but
the guides assured us there were four fine streams.
The location usually given to the Chippewa applies
well to the Harmony, and it may be they are the
same river under different names. Our ordinary
maps of the northern shore of Lake Superior are
altogether imperfect, and even the charts of the
Hudson's Bay Company are not entirely accurate.

Anxious to explore the stream, no sooner was our
camp pitched and dinner over than we embarked
and continued the ascent, being poled against the
current by the two guides, and trying every promising spot as we passed. Fish, however, were nowhere to be found, and disgusted with the heat
that not only annoyed ourselves but had destroyed
our sport, we were about giving up, when Frank
stopped the boat over against the mouth of a little
murmuring tributary brook. There were a quantity
of small stones and large rocks where the rivulet

joined the river, and the cast being a long one, I extended my line and dropped the fly just where the two currents met. It was taken instantly by a fish that, after fifteen minutes' vigorous play, was landed and found to weigh two and a half pounds.

That inaugurated our sport, and was followed by the capture of at least two dozen magnificent trout, that were not only immense in size, averaging nearly three pounds, but were extremely beautiful and uncommonly vigorous. Their tints were rich and dark, differing as greatly from the lake fish as the trout of the Canadian rivers differ from those of the salt water. They fought with great courage and perseverance, requiring skill and patience to land; and anxious as we were to take a large one, that is to say, one of over four pounds, those of two and three pounds were so numerous and voracious that we could not effect our object.

We landed some by hand and threw many back into the water, but, notwithstanding, soon had more than we could possibly use. There being no reason for our taking any more, and Don having complained that the cast was inconveniently long on account of the imperfections of his rod, I assured him I could cast entirely across the pool, and to prove it, lengthened my line, and at the first cast hooked fast in the rock beyond. Not caring to break the line, we dropped the boat across the stream, and while passing over the pool, beheld the bottom literally black with fish. If we had been inclined to wanton destruction, we could doubtless

have killed a hundred; but having no means to pot
or souse them, and knowing that they are comparatively worthless salted or smoked, we had resolved not to kill more than we could eat.

On the way back to camp we took a long, lean,
poor, sickly fish, that, if in good order, would have
reached six pounds, but in its unhealthy state only
weighed two and a half.

At supper that evening Don made a formal protest and complaint, insisting that he would drink no
more tea till he had white sugar; he entered at
some length into the characteristics and peculiarities
of sugar in its various stages, questioned the advantage of using brown sugar at all, intimated that white
was the best, most economical, and least bulky,
advised me in future to take none other, and finally
having disposed of every conceivable case but his
own, inquired why, when we had abundance of
both, he was not allowed the one he preferred, by
which time I had it out and ready at his hand. He had
evidently braced himself for a terrible argument,
seemed somewhat surprised at the want of opposition,
and after a moment or two began to call in question
the propriety of opening a new package, when the
brown sugar was already in use; that, in fact, although some people preferred white, and he must
confess he was among the number, others liked the
flavor of the dark colored; that little inconveniences
were the natural concomitants of a sportsman's life;
that when a number of bundles were opened they
were more exposed to dampness—a serious injury to

sugar—and there were more packages to look after, and that he was decidedly of opinion it was unadvisable, and that he was entirely willing to go without his tea. By this time the tea was drunk and supper ended.

It is a delightful thing of a cool summer evening to sit round a rousing fire that casts its variable glare upon the trunks and lower branches of the stalwart trees, and gives a ruddy glow to the white tent, the dense underbrush, and the kindly faces of the honest guides. At such times, while listening to wild stories of woodsman's life, that are doubly interesting when repeated upon the ground where they occurred, a pipe is absolutely delicious. Every member of the temporary household selects a rock, or log, fashions a seat to his satisfaction as best he may, and recalls the events of other similar expeditions for the edification of his associates. On such occasions cigars, which are cumbersome at all times, do not seem to answer, and recourse is had to the little pouch of Killikinnick which every one carries with him; under the joint influence of story and tobacco, the time passes quickly away, and the hour of bedtime arrives too soon.

Notwithstanding the summer evenings are usually cool above the line of the British Provinces, we happened to have fallen upon a hot spell; and although the fire was not disagreeable, the mosquitoes, which are benumbed by cold, were lively and plentiful. Under these circumstances our mode of proceeding was to close the tent and then with a

candle carefully burn them one after another. To
do this successfully requires nerve and skill; the light
must be approached quickly enough to catch the
nimble fellows, and just far enough not to scorch the
tent; the operation gave Don decided pleasure,
especially as they are consumed with a loud "pop."
In course of the proceeding he incidentally remarked: "Their galleys burn; why not their cities,
too?"

Next day we ascended the river to the falls, which
were about three miles from camp, and were found
to be attractive neither to the fisherman nor the
lover of nature. The water was warm and fishless,
the shute was small and unromantic. We dined at
its foot, and descending, fished the pool that the
day before had rewarded us so satisfactorily. Our
prey was still there, eager as ever for hook and feathers, and soon covered the bottom of our boat
with their glistening forms. My line after some
time happening to become fouled in the bottom, and
skilful fishing appearing to be out of place, I laid
down the fly-rod, and taking the bass-rod, cast the
trolling-spoon with some effort and a loud splash
into the pool; instead of alarming the fish, it was
eagerly seized, and I kept on catching fish with it at
every cast, till Don became disgusted with such
unsportsmanlike procedure, and insisted upon returning to camp.

That day was made remarkable by the advent of
a thunder-storm, a rarity in the northern clime,
and the only one that occurred during our entire

trip. It was not violent, and had none of those terrible characteristics of similar phenomena in southern latitudes, and even in our regions would have been considered a tame affair.

As, however, it drove us within the tent, and gave us a little unemployed leisure, my attention was attracted to Don's baggage, which consisted of an incongruous assortment that would hardly have been thought of by any other amateur backwoodsman, and would certainly have astounded a professional. Of course there were abundant clothes of various colors and kinds, of which a buckskin under-jacket suitable for severe winter weather, but hardly necessary in a summer-trip, and a handsome dressing-gown, were prominent articles; also his shaving materials, very neat and elegant, that were not used till he returned; a thermometer that kept us informed as to the amount of suffering we were entitled to feel from the condition of the weather; a picture of his two extremely pretty children, set in a *passe-partout* frame, with a glass over it that was in daily danger of destruction, a bundle of toothpicks that would have lasted us both a year, a new and effective patent portable boot-jack, a clothes-brush and whisp, a bottle of *eau de cologne*, a pair of flesh-brushes, and many other things that might be classed as "odds and ends."

Most of these articles were jumbled together in a large water-proof bag, from which he was never known to be able to obtain any specific article without emptying the whole on the floor; but the picture, his looking-glass, comb, hair-brush, and soap he kept among the eggs. The eggs suffered considerably from the association, and their injury was felt by myself as head cook; but Don could never be persuaded to change his habits, producing abundant arguments to prove that that was their only appropriate place.

At supper he announced his firm conviction that china cups and plates were a necessity to existence, that tin was an abomination, and that on all future trips he should be properly supplied. He was indignant at a suggestion that they might be broken, and burst forth :

"You are so set in your ways that you think no one can have any ideas but yourself, or make any improvement on your plans. Here you are, drinking high-priced tea, and even brandy-and-water, out of tin cups that hold a quart,"—this was an exaggeration, as they were only pints—"have a disgusting taste that absolutely destroys the flavor, and are of such a shape that you have to dip your nose into the fluid before you can swallow any of it. With hot tea this is painful, and with brandy, or even water, far from pleasant."

"Glass or china would be more agreeable on some accounts——" was the mild reply.

"I should think so," he interrupted. "Allow me to ask what you paid for this tea?"

"One dollar and fifteen cents a pound."

"And what does it taste like?"

"Tea."

"Tea ! Well, there are some people that can hardly tell wash-basin slops from the best Bohea."

" But, then," I hurriedly explained, to moderate his disgust, "china is so liable to be broken ; I had once an entire case of liquor smashed by my guides."

" Yes, and that liquor-case is a case in point ; because that was lost you do not give up carrying liquor, do you ? Then why cease using china cups, not that they have been, but only from fear that they may be broken ?"

"They are so much heavier than tin," I remonstrated.

" As if the weight of two cups, one for you and one for me, and two plates, was so serious. Let's dispense with something else ; take less to eat, if you please, but have it decently served."

Convinced by this eloquence, I meekly promised to comply on our next expedition, but Don was not altogether satisfied, and continued :

"I do not wish you to consent to these views merely to suit my wishes. I want you to be convinced. I dare say there are advantages about tin ; it may be knocked about, is always ready at hand, is light, and stores in small compass ; for rough travel, doubtless, it is admirable, and, were we to make long portages, would be better than china. After all, the taste of tin must be more apparent than real ; the metal cannot come off, or it would dissolve ; and how, then, can it give a taste ? The pots are large, but a man wants a good, long drink, whether of tea or brandy, when exhausted with hard work or exposure. After all, you will find many advantages in tin cups, and, really, the plates are scarcely objectionable; before deciding, you must look at these matters from both points of view. However, as we cannot obtain china this trip, and as we are discussing improvements, there is one thing I insist upon hereafter—we must have table-cloths and napkins."

"What!" I exclaimed, absolutely overcome at this suggestion.

"Table-cloths and napkins. You have probably heard of such things before; they are customary at a gentleman's table, and if a person does sleep in a tent, he need not forget he is a gentleman. Look at this table, made out of two rough boards that were never even planed, transported in the bottom of our boat, and walked over daily with dirty shoes and occasionally with bare feet, sullied with the marks of promiscuous bundles, half covered with grease, and stained with tea, bilge-water, and fish-blood gracefully intermingled."

"That is too bad; they are two good, clean boards that Frank washes regularly, and which are in themselves an unusual luxury; for in wood's-life we usually dine off a log or a flat rock."

"They may be washed occasionally; but as dead fish are first gutted on them, and as tea and grease are afterwards spilled on them till they are revolting with filth, I do not see, for my part, how you can eat your dinner off them."

"I don't eat off them; I eat off my plate."

"That you may call a joke; but hereafter I shall

have table-cloths and napkins. You carry towels, why not napkins?"

"Because you cannot stow a large number, and if you have only a few, how are they to be kept clean? The guides have enough to do without trying to wash table-cloths with cold water and no starch."

"If that is so, I should take an extra man to wash them."

The next day we met with a loss. We had noticed that the Indians, when they travelled, were invariably accompanied by their dogs; these were rarely accommodated on board the canoes, and followed along the shore, swimming the inlets or crossing at the head, making often much longer journeys than their masters, who passed from headland to headland, but coming up with the camp at night to partake of the frugal meal. Sometimes, however, they strayed, and either lived on chance gleanings from travellers or perished in the woods. There were two ownerless dogs near our camp, and although precautions had been taken by our men, they succeeded in carrying off our only ham, leaving us nothing to show for it but the empty bag.

Don's appetite had been sharpened by open air and exercise, and he expatiated at length upon disappointed hopes of fried ham, broiled ham, ham omelets, ham plain, and ham and eggs, and suggested many new and doubtless excellent dishes, of which ham was to be the principal part. His advice was valuable, but somewhat late.

Being already tired of the to me uninteresting
Batchawaung and its one pool of numberless trout,
and having a strong and favorable breeze, we broke
up camp, descended the river, killing a duck on the
way, and once out in the open water, headed for
the Point of Mamainse, which is Chippewa for sturgeon. The wind, however, soon came out ahead,
increased to a gale, and drove us into *L'anse aux
crêpes*, or Pancake Bay, where we were detained
that day and night.

L'anse aux crêpes is at the mouth of a little rivulet that tumbles over scattered boulders, and occasionally contains some nice trout; but the water
was low, and although we caught enough small fish
for supper, we did better with young ducks, happening to get a shot into a brood, and killing with
the two discharges seven plump, luscious, well-
grown little fellows, which replenished the gridiron
finely.

The temperature fell to thirty-seven degrees, and
with it the mosquitoes—a delightful change from
the oppressive heat and hungry hordes that had tor-
mented us. We camped for the night at the mouth
of the rivulet, and continuing our voyage early next
morning, soon reached the bold, imposing promontory called by the Indian name Mamainse. The
shore is rocky and precipitous to such an extent,
that the fisherman finds difficulty in casting the fly,
or even pursuing his way along the steep cliffs.

The water is filled with broken rocks, as at other
parts of the coast, and where these project above

the surface a good stand is obtained. At one spot the waves had worn out a deep cavern, where a dozen men could sleep, protected from the air, and often under foot could be heard the smothered rumbling of the water as it rushed into deep holes out of sight. Above the bare rocks, which are often fifty feet perpendicular, stretch the sparse underbrush, the stunted evergreens, and the moss-covered granite of the mountains, till they reach an elevation of a thousand feet. Frowning down upon the water stands the Point of Mamainse, a rallying-spot for the summer fogs and winter storms, a landmark to the voyageur, a barrier to the fiercest commotion of the lake, and the upper boundary of Tequamenon Bay, as the confined portion of Lake Superior near its outlet is called.

It is an extensive promontory, and point after point presented itself to our wearied eyes; we landed, rose, and lost some fine fish, and killed several of good size; but as the wind was adverse, we could not afford to waste time, and pursued our journey till nightfall.

Next morning we tasted a Batchawaung trout that Frank had salted and smoked by hanging near the fire; inasmuch as it was green and had not lost its original flavor altogether, it was quite appetizing; but a smoked trout that has been dried sufficiently to keep, is about as hard, unpalatable, and indigestible a morsel as man can put in his mouth. It has neither the flavor of the mackerel nor the richness of the cod, and not the slightest pretence to the delicacy of the salmon. Slightly salted and smoked, however, it will remain good for several weeks, and furnish a variety to the woodsman's Spartan fare.

Unfortunately there is no way of preserving trout; these fish, so delicate fresh, are almost worthless pickled, soused, salted, or smoked; while those of a size to be worth catching are too large to preserve by potting, in which way alone can their flavor be preserved. They are pickled by being immersed in water that has had sugar and salt boiled in it; they are soused by being cooked and preserved in vinegar and allspice; they are smoked by being salted for a night and hung in a smoke-house or near the fire; they are kippered by being rubbed with salt and a little pepper, and hung in the sun; they are potted by being cooked and packed tightly in jars, and having hot lard or butter with spices run in and over them. Only when prepared in the latter way are they eatable, and then only when they are small.

This day we had our first really favorable wind that bellied out our sail, and relieving the men from the labor at the oars, drove us along at a famous rate, enabling us to push boldly out into the lake that was alive with the dancing, foam-crested waves, and urging us onward famously in a direct course.

When far from shore and miles from the habitations of a civilized being, we espied approaching another barge similar to our own, and which proved also to be carrying a party of fishermen.

Our sail was hastily lowered, and the vessels being laid alongside of one another, we held an interesting conversation with our fellow-travellers. It appeared they had ascended the Neepigon, and gave glowing accounts of the number of fish, but not much of the character of the fishing; saying that the trout, which were large on the average, were collected in pools as we had found them in the Batchawaung, and were so numerous as to ruin the sport. They had had a long journey, and were out of whiskey, a deprivation that we hastened to supply; and were glad to see civilized beings, and to feel that they were once more on the confines of the land of the white man.

With mutual good wishes we bid them farewell, and watched their barge after we separated growing smaller and smaller in the distance, till it was lost to view. How suggestive are such meetings of individuals who have never encountered one another before, who form an acquaintance as it were in the wilderness, shut out from the rest of mankind, and, separated, never to meet in the wide world again; like a ray of sunshine through a storm-cloud, shining for an instant across the surrounding darkness, gone in a moment, and never to be re-illumined, leaving nothing behind but a pleasant memory! Not one of the persons in either boat will ever forget that meeting, and nevertheless no conceivable circumstances can bring them together again on the boundless waters of Lake Superior.

We reached the Agawa that night. The stream

was sluggish at its outlet, near which a change in its course had left a small pond in the sandy shore, and was not altogether inviting, with its shallow, discolored, heated current. It has a high reputation among those who have explored it, but flows into the lake in a commonplace manner. A neighboring swamp encouraged the growth of mosquitoes; and the black flies, which seemed to be of an unrecognized and indescribably vicious species, were annoying in the extreme. There was a small settlement of Indians near by, and hardly had we commenced pitching our camp, which had to be located some distance from shore on account of the pebbly beach, ere they appeared.

There was an old man, the embodiment of harmless idiotcy, who turned out to be a patriarch and not the fool he looked; two fine-looking, straight-featured young men; two boys, a little girl, and three dogs. The latter evidently belonged to the family, for they all, dogs included, stood in a row, the latter fully as intelligent as the former, and none of them offering the least assistance while our men and ourselves raised the tent. The old man wore a conciliatory expression of imbecility, the young men a confirmed air of vacuity, and the dogs and children seemed imbued with a few sparks of intellect.

They made no motion and uttered no word till a fire was lighted, when they instantly crouched round it. As a race, living in the rudest manner, and debased from their native simplicity by contact with the white man, they have small claims to intelligence; but to their credit, be it said, they are ordinarily honest, and unless grossly outraged, perfectly harmless.

They are readily moved to laughter, greatly enjoyed the appearance of our hats, which were stuck round with flies, and shouted with delight atthe noise made by Don's click reel, when he took a trout in the small pond previously mentioned, and throughout our intercourse with them, proved themselves pleasant, trustworthy companions.

While our guides were preparing supper, Don proceeded to explore the neighborhood, and made his way to the wigwams, where he found more of the same family. Immediately on our appearance, the women, after peering furtively through the chinks, retired into obscurity, ignorant, probably, of our high delicacy towards the female sex; and in fact throughout, betrayed a disgusting want of confidence; the three favorite wives of the silly old patriarch, wives that we were told were both young and pretty, having fled into the bush before our canoe had touched land. During our entire stay we had nothing but dissolving views of female charms—loveliness that was not arrayed in crinoline—although Don devoted every spare moment to persistent visits.

A young man appeared promptly from under the blanketed door of the first wigwam, and Don commenced an instructive conversation on the subject

of numerous dogs that were howling round in unpleasant proximity to our calves.

" You have a large number of dogs ?"

" Ya."

" I suppose you use them in the chase ?"

" Ya."

" They accompany you in your journeys ?"

" Ya."

" What do you chase with them ?"

" Ya."

" I asked what do you chase with them ?"

"Ya."

" Oh, I see you speak French."

" Ya."

" *Qu'est ce que l'on chasse avec les chiens ?*"

" Ya."

Don now began to doubt whether his new friend
spoke either French or English, and had recourse to
Chippewa, at least as near Chippewa as he could come.

" Vat you chase, chassy, vis the doggees ?"

" Ya."

" You chase *les cerfs* the deer, the elks, the moose ?"
gesticulating freely.

" Ya."

" The beaver, the—the—*castor?*"

" Ya."

" The rabbit, the—the—ze rabeet ?"

" Ya."

" Don," I burst forth at this stage, " he does
not understand a word you are saying."

" On the contrary, he evidently understands perfectly, how else could he answer so intelligently ; of course he does not pronounce yes accurately, but is entirely comprehensible."

" Well, then, ask him about the canoe he is building ; how many it will hold, what those strings are for, and where he caught that large trout yonder ?"

" You build ze canoe ?"

" Ya."

" How many it hold ?"

" Ya."

" It hold one ?"

" Ya."

" It hold two ?"

" Ya."

" You see he says it holds one or two."

" Well, now about the strings."

" Zese strings, what for ?"

" Ya."

" No no ; what for yese strings ?'

" Ya."

" What zay use for ?" raising his voice.

" Ya."

" You no understand ; what for, what for ?"

" Ya."

" Leave the strings and try the fish ?"

" You see ze trout, *truite ?*"

" Ya."

" Where you catch him ?"

" Ya."

" Up ze river ?"

" Ya."

"Or near by ?"

"Ya."

"No, no ; where catch him ?"

"Ya."

"Here or zere ; here or zere ?" very loud, as though the savage were deaf.

"Ya."

"That will do ; and after this instructive conversation we had better seek our camp and supper."

"Just as you say; he evidently does not fully understand the last question, although I think we might obtain some valuable information from him. We certainly want to know where he took that fish, which must weigh four pounds."

"We certainly shall not find out, as baby talk evidently is not Chippewa, although I wish it was, and will need Frank's aid in our communications."

The other Indians were still seated near our fire, and received with apparent thankfulness the remnants of our supper, of which we took care that the little girl should have her share, after we had finished. As the river was low and could not be ascended with our barge, nor without much labor on foot, it was necessary to hire canoes; but unfortunately we had nothing but United States money, which was about as worthless as white paper. Frank took ground that we should pay them in stores of pork and biscuit; but as he seemed utterly regardless of our anxiety to make a positive bargain, and but little mindful whether they were paid or not, Don felt it necessary to approach the subject cautiously, and having read of the pipe of peace, thought the opportunity a good one for its introduction. Taking out his pouch, he gave them enough tobacco to fill their pipes all round, having learnt from Frank that it was not necessary to pass his own from mouth to mouth, which he had considered imperative, but which was not altogether pleasant. He was solicitous about their having their pipes well lighted, and being pleased with the tobacco, and when reässured on that head, and satisfied that genial smoke was producing its natural effect, he permitted Frank to give a few gentle hints suggestive of our desires to ascend the river, our possession of quantities of pork that we did not wish to take back with us, and our anxiety to be satisfied that canoes could be had.

The subject being skilfully launched, Don expressed great interest in the little girl, whose name he found was *Wajack*, which being interpreted, means Little Rat, and finally made his great point by the production of his picture. This had hung in our tent night after night, had been carried in our basket day by day, and had smashed its score of eggs; but now it repaid us. The hearts of the savages were won, their delight was rapturous, expressions of admiration were universal, the highest encomiums were passed upon it, and the little children, whose likenesses were really extremely pretty, were as the perfection of loveliness as Frank interpreted it, pronounced to be " so nice and fat."

5

This we felt to be our moment of victory, and Frank was directed to improve it. Standing before the fire, with a gridiron in one hand and a dish-cloth in the other, he burst into a strain of unequalled eloquence. Without understanding a word, we could imagine him painting our desolate condition; how we were strangers from a far-off land, had left the pale-faces, our wives, our little ones, bringing with us only their faint delineation on paper, in order that we might see the beauties and grandeur of the Indian's home—to sleep in the woods, to float upon the lakes, to wander through the forests, to explore the rivers. How we felt the red men to be our brothers, and wished to know them better, wished to stay long with them, to voyage in their company and under their guidance; that we were great men in our own land, but knew little of the wilderness or the manners of savage life; that we were rich in corn, in pork, in flour and biscuit, but had not thought to bring our purses, which were filled to overflowing, with us; but that we felt our brethren of the great Chippewa tribe would befriend us, would supply us with canoes and guides, and help us on our way. That the great universal brotherhood of man demanded it, and that the time might come when they would be in our land, penniless and ignorant, and might have to look to us for canoes and guides; and would be glad to remind us of the time they helped us up the Agawa.

At the end of every sentence and at every pause, the Indians all, big and little, broke in with a simultaneous m–m–m, a sort of grunt that became more vigorous as Frank became excited, and grew louder as his arguments grew stronger; till before he was through, the listener would have supposed that the entire party was suffering in the agony of what children know as the stomach-ache. The grunt was not in the least like the conventional humph, was uttered without opening the mouth, which would have been an excessive and unnecessary labor, and was capable of great expression. It began sympathetic, grew appreciative and confirmatory, and at last became wildly enthusiastic, evidently taking its origin from the Greek chorus, which is of a similar appropriateness; it was the strangest accompaniment to a public speech we ever heard.

Feeling the importance of the case, we endeavored to keep our countenances; but what with Frank's bursts of eloquence, his graceful and impressive gestures with the gridiron, the vehement grunt in chorus at every pause, our strange position congregated in the wild woods round a fire with a parcel of unkempt savages, begging to swap off, as our Yankee brethren would say, a quantity of biscuit for a passage in a canoe, we could not contain ourselves, but rolled over in convulsions of laughter.

At first the Indians did not know what was the matter, then they joined with us, and when we attempted to imitate their grunt they shouted louder than we had done. Frank felt that aspersions were cast upon his eloquence, and seemed to have his

feelings hurt, but unable to resist the general hilarity,
at last joined the

"roar

That echoed along the shore."

What Frank had really said I never could find
out, but believe that he mentioned the subject we
had at heart no farther than merely to order the
young men to bring their canoes. Although half-
breed himself, he was influenced by the general
contempt for the rights of a savage, and determined
in his own mind to have the canoes and pay for
them as he pleased. Doubtless also he was more or
less controlled by a dread of self-depreciation in
acknowledging that he served penniless employers.
To our persistent questions he would respond laco-
nically that it was arranged, but would say nothing
as to particulars. As we were entirely in his hands,
having discovered that not a word of our language
did the Indians understand nor we a word of theirs;
and as, although our desire to do justice was great
and might have been strong enough to induce us to
give up the idea of obtaining the canoes, we were
utterly unable to communicate it, we were compelled to submit to Frank's course.

The Chippewa language is beautiful, easy, flowing,
graceful, full of vowels, expressive, capable of
vigorous impression, and, were it more generally understood, pleasant to acquire; but above all is it
advantageous when an entire ignorance of its mean-
ing enables you to take what you want and pay for

it as you please. And if the native is dissatisfied he cannot vituperate or abuse you, as the strongest word, *le plus vilain mot*, as Frank expressed it, fortunately is " *chien.*"

MOUNT KINEO.

CHAPTER V.

The canoes arrived on the following morning ere our breakfast was dispatched, and having stowed into them our fishing-gear and the requisites for a simple meal, we were about embarking when Don, who was directed to sit on the bottom of one, between the two Indian boys, entered a violent protest, and seating himself on a log instead, announced he should either not go at all, or should be allowed to pole and have sole charge of one end of the canoe. This proposition astounded all who could understand, and would have astounded the others still more if they had understood it; but ere we had recovered our breath Don commenced explaining his views :

"For many years I have heard of voyaging in a canoe; have thought it the chief pleasure of the wilderness, and have been anxious not only to learn how, but to do it. Of course, you will hardly expect me to know how to manage so frail a boat without practice, and yet if I never practise, how am I to learn? It is self-evident I must commence some time. If you admit that, and you can scarcely dispute it, what better time could I have than the present? You propose to take the bow of the other canoe, and although you are probably not as expert as the savages, you did not acquire such skill as you possess intuitively, but by experience. You will probably suggest that I may upset; if so, the consequences fall only on myself. You have put no stores in this canoe, and the ducking will be mine. Let one of the Indians stay behind, for I have counted upon this as my greatest pleasure."

"But, Don," I reasoned mildly, somewhat appalled at the prospective consequences, "you will smash the canoe."

"Oh, no; you did not do so when you commenced; and if I do, it is not worth over fifteen dollars, and I can pay for it. We have stores enough, and I can make up the difference to you."

"But you will never succeed——"

"Pooh, pooh! You succeeded, why not I? I do not ask you to give up the pleasure which I see plainly you are bent upon, but we can leave one of the Indians here; I will go with the other, and you with Frank. That will make the load lighter, besides."

"Has *monsieur* ever poled a canoe?" asked Frank, wonderingly.

"No; but I must commence. Of course, I will have difficulty at first, but it will come; do not trouble yourself about me."

"The work of poling against a strong current is tremendous, and the river being low, the rapids are unusually heavy. You will be entirely exhausted ere you have gone half-way."

"Do not worry yourself about my sufferings; although your argument is evidently defective, as low water cannot be stronger than high, if I fail to keep up with you I can lag behind or come home."

"Really, you do not know what you are undertaking; but I will tell you what you can do. Go with the two Indians, see how they manage in the first rapid, and then take the place of one and try it."

To this, after much protest and complaint, Frank and I persuaded him to agree ; more, however, as a personal favor to ourselves than on any other ground, and his grumblings of dissatisfaction were loudly audible till we had passed the first rapid ; Don neither offered to pole nor grumble afterwards.

The water was very strong, collected in large pools, and then rushing with tremendous force down a confined channel, or else pouring in long exhausting stretches of foaming current over pebbly shallows and amid protruding boulders. At one spot Frank and myself were fifteen minutes, just able to hold our own and not advancing a foot, with the imminent risk of upsetting at any instant ; and when I was out of the canoe fishing, he was utterly unable, to the intense delight of the Indians, to stem the rapids at all.

The canoes were small, and the canoe-men had to occupy a most uncomfortable position : kneeling and sitting on their heels, not being able to stand erect as I had often done in larger boats, so that Frank complained of cramp in his legs for days afterwards. Short setting poles were used, and our utmost strength had to be exerted where the current was strong. Of course, the Indians were entirely at home at the work, and although straining their best, enjoyed our deficiencies and shouted over our mishaps; whenever we either caught a trout or came near upsetting our canoe, whenever we had any good luck or any bad luck, and often when we had neither, they roared with laughter. Not appearing to give the fate of their canoe, which was in our hands, a thought, they were intensely amused whenever we brushed against a rock or careened her till the water flowed in. Instead of the proverbial taciturn grimness of the conventional Indian, they were hilarious and loquacious, although their language was a sealed book to us. They were on the best footing, and held animated conversations with our guides, were continually amused at their own witticisms, and when on our return, while descending an unusually dangerous rapid, Frank, distrustful of my judgment, insisted upon taking entire charge of the canoe, and as a natural consequence came very near upsetting and throwing us into the boiling waters, to the peril of our lives and destruction of the boat, they could hardly contain themselves, but made merry over it the entire way home.

The Agawa winds among high, bleak, and sterile hills, is rapid and filled with pools, but has none of those tumbling cascades which give life to the water and wear out deep, dark holes where trout love to congregate in warm weather. The current, stained with the dead leaves and decaying vegetation of the ponds and marshes, where it has its source, is amber-colored, and lends its hue to the pebbly bottom over which it flows. It evidently, throughout its great extent, furnishes admirable spawning-grounds for the fastidious trout, and in cool weather is filled with them in vast numbers. But when a warm season has heated the water, and a drouth has diminished the current, the fish, finding the element unsuited to their comfort or even existence, are compelled to seek the cool, shady caverns of the lake shore.

The river, when we visited it, was in this condition, and there were none but small, dark-colored fish, which, although excellent in the frying-pan, after the excessive exertion of surmounting the rapids had given us an appetite, furnished but tame sport on the line.

Our dinner was pleasant, our trip exciting, the scenery wild, the river interesting, the savages amusing, and ourselves agreeably entertained ; but we returned early, possessed of a wretched show of game. We had taken two dozen fish, but none of them were large.

On issuing from the secluded channel of the river, we realized, to our surprise, that a heavy gale was blowing from the south-east. We had not felt the wind till we approached the open water, and emerged from among the hills and trees, but soon found the waves rolling in upon the sand-beach in a way to remind one of the surf on " Old Long Island's sea-girt shore."

The waves appeared to drive the trout in from

the lake, and towards evening the river near its mouth was alive with them, breaking in every direction ; yet, strange to say, although we cast our flies frequently directly over them, and kept on fishing till it was night, not a trout did we take. In all our experience such a thing had never happened, and where they were so numerous, a dozen often being visible at the same instant, so voracious and unaccustomed to the presence of man, it was extraordinary. Fish will frequently, although breaking freely, refuse the fly, but generally a few will be misled, and occasionally one will be caught ; but here in the Agawa, a hundred miles from civilization, we saw ten thousand trout in the space of five hundred yards, and after expending skill and patience, failed to take a single one.

No explanation of this phenomenon presented itself; there was nothing in the air, water, or time of day to explain it, and although it was followed during the night by a great change of temperature, there would appear to be no connection between the two events. The fish seemed to be playing rather than feeding like salmon running in from the sea; and, anticipating cooler weather, may have been preparing to ascend the river. And it is proper to mention here that two gentlemen, who fished the river a few weeks afterwards, had remarkably fine sport.

Fishing having proved itself vanity and flies a misconception, we returned to the tent and superintended the payment of the guides, by impressing upon Frank the necessity of giving them sufficient. One received his in a greasy, dirty hat that he had worn for several seasons, and which could hardly have improved the flavor; and the other, not having so expensive a luxury as a hat, wrapped his in a neck-cloth that had been in use day and night for years, and had never been washed. Frank gave them each, in addition, a little butter on a biscuit, and they hurried away, delighted with their treasures.

The Indian children had brought a number of agates that they had collected from time to time, and Don selected the best, which were, however, inferior specimens, and paid for them also by barter. Of course, our little friend Wajack had her store to exhibit, and received a favorable consideration from Don, who endeavored to make her understand a few English words, which were such exquisite baby-talk as to be nearly incomprehensible to the rest of us. He found in the long run that he succeeded better by holding up the proposed payment and pointing to the agate, as none of the savages presumed to ask for more than we offered.

The following morning the trout again declined positively to recognize our allurements, and the wind being fair, we concluded to commence our homeward voyage. We were sorry to part with our amusing Indian friends, notwithstanding an occasional pang of fear for our numerous articles that lay scattered about, and which it is only justice to say were entirely untouched · but as we could make nothing of the fishing, had become possessed of the best agates, and had explored the river thoroughly, we proceeded to reëmbark.

The wind was, for the first time, in every way favorable; but ere we had reached *Point aux Mines* it had become so violent that Frank, alarmed at the increasing *roulan,* began to talk of his wife and eight children, and how sorry they would be if he were drowned; and when the wind further increased, and Frank began to talk of his nine children, we concluded it was time to stop and put into a port of distress. In truth, those open, heavily laden boats are not the safest of vessels in a seaway, and yawing about as they do before every wave, have to be watched carefully lest they broach to and fill.

Charley enjoyed Frank's terror, and would have kept on as a matter of pride till his employers were satisfied; but Frank, with streaming hair, staring eyes, and blanched countenance, was a picture of distress, and if we had not given permission, would have taken it to run behind the first friendly point.

This proved to be *Point aux Mines,* where in former days a copper mine had been located, and the shafts and buildings, dilapidated it is true, and fast crumbling to pieces, remained to mark the traces of man's enterprise. The point had been purchased by a company from the Crown; but as the latter failed to pay the Indians, who were the rightful owners, they, with the assistance of many of the Canadians, among whom was our friend

Charley, made a night-attack upon the post, and, by a complete surprise, captured it without loss or bloodshed. The speculation never having been profitable, the company was only too glad to be captured; and having obtained an extravagant indemnity from the home government, never resumed possession of the works.

The buildings were windowless and tenantless, and served as shelter for voyaging parties of Indians; the underground passages were falling in, the machinery was going to ruin, the platforms were rotting, and the gardens had grown up with long, rank grass.

We explored the shafts, collected some specimens of the ore, and returned to the boat in time to find the wind greatly abated, and embarking, soon arrived at the Point of Mamainse. Having fished for a short time from a rock named after one of our best New York fishermen, Stevens's Rock, we continued our voyage, and reached the former camping-ground on the Batchawaung before dark.

The weather had changed. The rain was falling in that dull, penetrating drizzle that is so depressing to one's spirits, and the cold air made our wet clothes and damp bed far from comfortable. Camping in a rain, building a smoky fire from damp logs, and making a bed of wet boughs, in spite of the protection of water-proof blankets, is unpleasant, although it rarely produces sickness. Don bore the discomfort with a patient composure that was an eminent example to our city exquisites, and never uttered a complaint; on the slightest provocation he would probably have proved, conclusively, that moisture was man's natural condition, and infinitely preferable to sunshine and dry clothes.

On ascending the river next day, as Don and myself were walking along the bank we observed a rustling in the grass, and pausing, roused a flock of partridges. I shot one as they rose, and beholding them, to my great satisfaction, alight on the neighboring trees, proceeded to poach, thinking only of the pot, and shot from the trees and on the ground, in utter disregard of all sportsmanlike rules, the entire covey. They consisted of but a single brood, and the young were not more than three-quarters grown; but the anticipation of their juicy tenderness on the gridiron overpowered any qualmish sentimentality, and right glad were we to collect the ten plump, tender little fellows into a bloody pile.

The trout had moved from their former locality, but were plentiful as ever, enabling us to satisfy our desires and return early to camp, with one fish of four pounds and several of three. During the day there was a sudden change of temperature, preceded by a furious attack from the brulots upon our unhappy persons. Apparently anticipating the advent of cold weather and partial lethargy, they satiated their appetites with our blood, in spite of ointments and veils.

During our absence a party of fishermen had arrived from the Sault, and finding our camp, located themselves a few hundred yards below us. As we descended the river next morning, we stopped to exchange salutations and inform them of the condition of the fishing. Being ourselves abundantly satisfied with killing trout, we proposed making a short visit to the romantic Harmony before returning to the Sault, and left the strangers in the sole possession of the Batchawaung.

We found the Harmony lower and warmer than we had left it, almost deserted by trout, but otherwise as beautiful and picturesque as ever. We lingered round the falls, and listened to the noisy cascade, drank from the ice-cold spring, shot a few ducks on the lower stretch of water, killed a dozen fine trout at the upper *shute*, and indulged in the luxury of laziness.

Don had been heretofore as active as any member of the party, often up the first and to bed the last ; frequently rousing the guides from their slumbers by a loon-like call, repeated until they appeared ; but on our first morning at the Harmony he positively refused to get up, and to my persistent entreaties, replied in a despondent voice :

"It is no use ; you give me no rest, keep me up every night till eleven, work me to death all day, and let the flies and mosquitoes annoy me without cessation. I will stand it no longer, and intend to sleep as late as I please."

"But, Don, breakfast is ready, and you will lose it."

"Then I shall have a second breakfast. You feed

me on pork, and trout, and ducks, till I am tired of
them, and get no nourishment from the endless repetition."

"I have made a beautiful omelet this morning,
and it will be ruined."

"Then make me another—we have plenty of
eggs—or I will make it for myself."

"But you will miss the morning's fishing."

"I do not care. I have caught trout enough to
last my lifetime, and I will have a little rest."

With that he turned over, incontinently went to
sleep, and no efforts on our parts, nor shouts from
the guides, who with delight imitated the cry with
which he had been accustomed to wake them,
could rouse him till eleven o'clock. Apparently
much refreshed, he eat a light lunch preparatory to
a more substantial dinner, the hour for which had
almost arrived. Getting up at eleven o'clock in the
woods is equivalent to sleeping till four in the afternoon in the city.

Somewhat moved by his complaints, and having
plenty of leisure-time, I devoted myself to providing
for dinner the best our larder afforded : soup made
from preserved vegetables furnishing the first
course ; trout, larded and fried, the second ; broiled
duck, garnished with thin pieces of pork, the third ;
and such entremets as boiled rice, chow-chow, and
the like, closing with a dessert of that remarkable
and ill-named preparation called corn-starch, one of
the most valuable discoveries for the city-bred explorer of the woods.

Corn-starch is a remarkable edible, supplying the greatest variety possible, never seeming to result in the same production, and furnishing a subject of untiring wonder as to what form it will take next. On some days it would be beautiful, transparent, bluish jelly, then it would be a solid, opaque white, and again a dusky brown semi-liquid substance; frequently it resembled pap, and now and then would be full of doughy lumps, as though endeavoring to effect an experimental pot-pie; sometimes it tasted of liquorice, at others it seemed flavored with molasses; but generally it had not the slightest particle of taste. I never could calculate on a result; if I tried to obtain jelly, I made pap; if pap was my purpose, pot-pie would be the product.

Don eat it daily in a state of bewilderment bordering on idiocy, inquiring regularly after the first taste: "What have we here, now?" But once, when brown instead of white sugar was used, and effectually obliterated all other flavor, he made what young ladies call a face. The inventor of corn-starch must be a wonderful man, but it is to be desired that he would reduce his bantling to a little better state of subjection, and put on his labels directions more applicable to the woods, where milk and moulds and flavoring extracts are not to be had, and ice-creams are a reminiscence of the past.

Monotony is the drawback to life in the woods, and corn-starch is doubly welcome on that account. It is nutritious, being composed of the essential portions of the grain, is compact, and easily protected from wet; it furnishes an astonishing variety of desserts where any dessert is a luxury, and it is an admirable addition to one's stores, though I wish it had a little more taste.

The dinner, including the corn-starch dessert, was a success, and revived Don's spirits, so that he was up betimes thereafter during our stay at the Harmony.

With reluctance we bade farewell to the pretty stream, whose soothing murmurs, grateful shade, and wild scenery invited us to remain; and our eyes lingered on the hills from which it springs, as we slowly passed out of Batchawaung Bay on the route to Gros Cap and the Sault. But, aware that our limited time was almost expired, we pushed on our homeward way, stopping to dine at the campground near its mouth. Here we found, amid the débris of ancient wigwams, the bleached skulls of numerous beavers, and were surprised at the peculiar formation of their long, mordant teeth. We had frequently noticed logs of considerable diameter that had been cut through by these powerful natural saws, and that bore the long furrows that they made; but were astonished to find, in extracting these teeth from the skull, that they constituted nearly a semicircle. Worn as they would be by severe and continued use, nature had made this provision to supply the rapid waste, and the portion of the ivory concealed in the skull was fully two inches long. Don collected several, and finding a peculiarly large specimen, muttered, on withdrawing the teeth, that it must be the remnants of

"Ahmeek, the king of beavers."

Before reaching Gros Cap we struck and lost, by the fouling of our trolling lines, which were both out together, a very large lake trout. This fish, in spite of his size, gave so little play that we were scarcely aware that we had hooked him, and were astonished when we saw his immense proportions as he came near the boat. We scarcely considered his loss a disappointment.

We spent two days at Gros Cap, having fine sport and killing some large fish. Don broke his tackle several times, and the lively, bright-colored, vigorous trout, luxuriating in their appropriate element, the cold spring water of the lake, gave us excellent play. Wandering from rock to rock, and casting out into the limitless lake, every rise was sudden and unexpected, every step changed the distance of our cast and the character of the fishing-ground.

The submerged rocks were visible through the limpid water, and from beside them or from their deep, dark fissures a trout might rise with a furious, impetuous plunge at any moment. The fish were numerous, breaking in the placid evenings in myriads, and the sport was entrancing. During the warm mid-days, when the sun was too brilliant or the lake too calm for fishing, we would wander about the island, hunting specimens, inspecting natural peculiarities, and chasing the *ephemeræ* that had supplied the place of the brown *phryganidæ.*

There was a surprising similarity of color in all the natural flies of that region; they were mostly of modified shades of brownish yellow or gray. The yellowish variety had two long whisks, one inch and three-quarters long, banded with gray, eyes round, white, and protuberant, with a black speck, and eight sections to the body. They were quite active and numerous, while other varieties resembled them in general appearance and characteristics.

The rocks were seamed with veins of copper, the oxide of which had discolored the adjoining stone, and occasionally we could obtain pretty and apparently rich specimens. Unfortunately, neither Don nor myself, though well enough read in the classics and other equally useful sciences, had ever studied mineralogy, and were as good judges of minerals as a savage would be of a watch. Our ignorant conclusions, however, were that if the north shore of Lake Superior were properly explored, under Yankee supervision, mines might be discovered equalling those of the south coast. With this sage conclusion we were forced to be satisfied.

Charley had a passion for prospecting; was ready at a moment's notice to dig out with the axe any strange-looking deposit, fully convinced that some day he should make his fortune, if he only could learn to distinguish the valuable from the worthless.

At last a strong westerly wind came out, and a

heavy fog settled down upon us, wrapping the hills in its graceful shroud, hanging pendant from the distant rocks and trees, shutting out the lake from view, covering the bushes with glittering gems, and wetting our thin clothes uncomfortably. As there was too much sea running to fish, we wrapped ourselves up in the water-proofs, and embarking the remnants of our property, set sail for the Sault.

This was to be our last day on the lake, our last day in the open woods, the last time we were to stand face to face with nature's solitude—and our spirits felt depressed at the prospect. No more sleeping beneath the cool canvas, no more looking out upon the limitless Big-Sea-Water, no more peering up into the silent night, and no more of those thronging thoughts and grateful inspirations that feed the soul in the wilderness. The freedom from rules and restraint was to be laid aside, the easy dress must be replaced by the methodical cut, the manners and acts must be shaped to those of others, and we were to conduct ourselves henceforward according to the received and established pattern. We were approaching civilization, where stiff and stately houses were to limit our views, and man's works shut out those of God.

The wind soon hauled ahead, and driving back the fog, let down a flood of sunlight on the sparkling water; but the current being quite strong in our favor as we approached the outlet, we made good headway, passing in our course a yacht crowded with sportsmen, and under full sail going wing and wing for the Neepigon, encountering other sailing vessels, and meeting with occasional evidences of man's presence.

At six o'clock that evening we shot the rapids, and discharging our load at the wharf, ensconced ourselves once more beneath the hospitable roof of the Chippewa House. Three glorious weeks had come and gone since we were last there—three weeks of unalloyed happiness, three weeks of invigorating life and exercise, worth all the medicines in the world—three weeks of intelligent and sensible enjoyment. In that time impressions had been made and lessons had been learned never to be forgotten; health had been acquired that would last for years, joy tasted that would leave its flavor during life. And now farewell to the staunch old barge; farewell to our canvas home, to the merry camp-fire, to the woodsman's life; farewell to the deep forests, the sombre pines, the waving elms, to the dancing streams, and the open water; farewell to our faithful guides; farewell to the graceful trout, the elegant namægoose, the fierce black bass; a long farewell to Gitche-Gume, Big-Sea-Water, the greatest of the great lakes of our great country!

CHAPTER VI.

THE finest trout-fishing in the world is to be ob tained at Lake Superior; although larger fish may be killed in the lakes and streams of Maine, and greater numbers in the brooks of New Hampshire, Vermont, New York, and Pennsylvania, nowhere is to be found the same abundance of trout, averaging above two pounds, and wonderfully game and vigorous, and nowhere a more beautiful region to explore or pleasanter waters to fish over. The entire rocky shore of the lake, along both coasts, is one extensive fishing-ground, where the skilful angler can at any point find delightful sport; the innumerable tributaries, large and small, of the British or American territory, unless shut out by precipitous falls, are crowded with myriads of the speckled beauties; and the rapids at the outlet furnish trout of the largest size.

The true mode of enjoying the sport is by camping out, when the adventurous sportsman roams from point to point and river to river, from camping-ground to camping-ground, at his own unrestrained will, varying the sights and sounds of beauty that are ever present in the wilderness; but excellent fishing can be had at numerous places;

united with comfortable accommodation. At the Sault St. Marie, at Marquette, at Grand Island, and at Bayfield public-houses are to be found, and so plentiful a supply of fine fish that the heart of man cannot fail to be satisfied; but the finest sport is to be realized along the Canadian shore, where camping-out is a necessity; for while on the southern coast the trout average a pound, on the northern they will run fully two pounds in weight.

To reach Lake Superior from the Eastern States the angler must either take the steamers at Cleveland upon days advertised in the local papers, or join them the next evening at Sarnia, by the Grand Trunk or Great Western railroads of Canada. He will reach the Sault in three days from Cleveland, and can save twenty-four hours in going by the way of Sarnia. At the Sault he will find unequalled bait-fishing, and occasionally excellent fly-fishing; but here, on account of the depth and strength of the water, the bait will kill the largest trout. At this thoroughly American village there is a well-kept hotel, the Chippewa House, and nearly all the requisites for camp-life, except the tent.

A few miles below the Sault the Garden River affords good sport and fair-sized trout, but is a difficult stream to ascend, while the first promontory on the southern shore of the lake, called White Fish Point, has long been famous as a fishing-station. At Marquette, which is a regular stopping-place for the steamers that traverse the lake, the waters are somewhat fished out; but about thirty miles to the eastward, within an easy day's sail, at Grand Island, there is splendid fishing, magnificent scenery, and a passable boarding-house. Here are the famous Pictured Rocks, ornamented with the fantastic hues of many-colored sandstone, and worn by waves and storms into a thousand odd shapes and strange resemblances, hollowed out into caverns, washed away into pinnacles and spires, at one place representing a yacht under full sail, at another a turreted castle of the olden time.

About sixty miles beyond Marquette are the Dead, the Yellow Dog, and Salmon Trout rivers, which are apt to be encumbered with drift-wood and underbrush, but which are filled with fish, and from one of which a brook-trout of six and a half pounds was taken. The photograph of this fish, or another of about the same size, is preserved at the Sault.

At Bayfield, the further terminus of the steamboat route, named after the first American explorer and surveyor of this region, is the best of fishing, united with good hotel life. In the neighborhood of this village two hundred and fifty pounds weight of speckled trout have been killed in one day by one good fisherman and one poor one ; fish of two and three pounds are common, and in the sheltered channels, between the Apostle Islands, the namægoose are taken in unlimited quantities. The Brulé River, and the many streams that empty into the lake in the neighborhood, although often choked with drift, are filled with fine trout.

On the north shore, amid the interminable forests that stretch in primeval solitude to the northern sea, enlivened only with the voice of the Peebiddy bird and one other melancholy warbler, beautified by a rare sprinkling of native wild-flowers,

"In the kingdom of Wabasso,
    In the land of the white rabbit,"

and along the Canadian shore of the lake, is the paradise of the fly-fisher. Every river swarms, every bay is a reservoir of magnificent fish that find their equals in size, courage, vigor, and beauty only in the salt waters of New Brunswick and Lower Canada. The entire coast is one long fishing-station, the rivers are stew-ponds, and the lake one vast preserve; at every step the angler may cast his fly into some eddy of the discolored stream or over some rocky shoal of the limpid lake with a fair prospect of alluring from the depths a glorious embodiment of piscatory power that shall struggle and fight, leaping from the water, and making many fierce rushes for a good twenty minutes, till he yields himself to the embrace of the net, exhibiting amid its brown folds the glorious silver brilliancy of the loveliest inhabitant of the liquid element. As he advances along the shore, an endless variety of water and land, continuous changes of rock and tree, and dark, bottomless depths or light gray shallows, present themselves to his eye; at one moment he is clambering along the steep, rough side of a precipice, whence he can scarcely toss his line

a dozen paces, at the next he is walking securely
upon some flat rock whence the receding hills permit him to cast to the utmost limit of his ability, or
he may ascend the nearest stream by the aid of his
strong barge, or in the light canoe, or else wading
waist deep against the rushing current, and there,
overshadowed by the hills and shrouded amid the
waving trees, he can visit pool after pool, try eddy
after eddy, till he and his men and the boat are
loaded, and satiety bids him rest.

Along the lake there is scarcely a choice of locality; from the sandy beach at Point aux Pins to the
outlet of the Pigeon River—the boundary of two
nationalities—at every point, in every cove, trout are
to be taken, and often in abundance ; but probably
the best as well as the most accessible spots are
Gros Cap and Mamainse. Of the rivers the most
famous is the Neepigon, where barrels of trout,
averaging four pounds, have been taken in one day ;
but the Batchawaung and the Agawa are nearly as
good, and within a more convenient distance, while
the Harmony is unequalled for wild and romantic
scenery.

The fish of Lake Superior excel those of the other
inland waters, either in flavor or game qualities, and
sometimes, as with trout, in both. The lake-trout
and white-fish bring a higher price in the Detroit
markets than those of Erie and Ontario, have a more
brilliant color and firmer flesh, and the trout infinitely surpass in appearance, strength, and endurance the dull, logy productions of the Umbagog or

Moosehead Lake. On taking the fly and experiencing the astonishing disappointment, they make one rush like their fellow-sufferers the salmon, and finding the pain clings to them, they leap with the energy of grilse with wild repetition, in the vain hope of shaking the tormenting barb from their lips. Nor do they resign themselves after a feeble struggle, but retain strength for many a rush when the ugly net is exhibited, often smashing tackle, carrying off leaders, and breaking tips in the course of the contest. Their colors are exquisitely delicate, their backs transparent mottled green, their sides of pearly whiteness, marked with brilliant carmine specks and faint blue spots, and their fins of the hue of clouded cream. Their flesh is flaky and rich, seamed with curd, and delicious to the hungry sportsman.

After having fished from Labrador to the Mississippi, and killed trout in every State where trout are to be killed, I am satisfied that the fishing of Lake Superior surpasses that of any other region on our continent, and is, as a natural consequence, the best in the world.

There are several remarkable peculiarities of scenery, among which are the pictured rocks and the sand dunes; and the sparkling lake, when stirred by a gentle breeze, is beautiful in the effulgence of the vertical summer sun; but the forests are gloomy and sombre, nearly impenetrable on account of fallen trees, and in the lower lands grown up with vast ferns, those evidences of the antiquity of our conti-

nent; so that the sportsman is mainly confined to
his canoe and the narrow strip of lake shore between
the beating waves and the impending hills. Beneath
his feet are the hard rocks, seamed with yellow veins
of copper, or wave-worn pebbles sparkling with a
hundred varying colors, only less beautiful than the
glistening fish that the skilful angler entices from
the lake and lands among them. From this narrow
strip he surveys the broad expanse of the Big-Sea-
Water, and dreams of the countless myriads that
rest in its liquid depths.

He travels with ease and comparative comfort;
in the commodious barge he stows the innumerable
articles that fill the measure of a sportsman's luxuries, including among them a roomy tent, appetizing
delicacies, abundant clothes, and whatever else fancy
dictates. With the barge, which, although twenty-
two feet long, is light and draws little water, he
ascends the larger streams; or he hires some passing Indian and his birch canoe, that wonderful
structure so beautifully and accurately described by
Hiawatha:

"Lay aside your cloak, O Birch-Tree,
Lay aside your white-skin wrapper,
For the summer-time is coming,
And the sun is warm in heaven,
And you need no white-skin wrapper.

Give me of your boughs, O Cedar,
Of your strong and pliant branches
My canoe to make more steady,
Make more strong and firm beneath me.

Give me of your roots, O Tamarack,
Of your fibrous roots, O Larch-Tree,
My canoe to bind together,
So to bind the ends together
That the water may not enter,
That the river may not wet me.

Give me of your balm, O Fir-Tree,
Of your balsam and your resin,
So to close the seams together
That the water may not enter,
That the river may not wet me.

Give me of your quills, O Hedgehog,
All your quills, O Kagh the hedgehog,
I will make a necklace of them,
Make a girdle for my beauty
And two stars to deck her bosom.

Thus the Birch Canoe was builded
In the valley by the river,
In the bosom of the forest,
And the forest's life was in it,
All its mystery and its magic,
All the lightness of the birch-tree,
All the toughness of the cedar,
All the larch's supple sinews;
And it floated on the river
Like a yellow leaf in Autumn,
Like a yellow water-lily."

And in this thing of life and beauty the fisherman
finds his way to the head waters of the smallest
brooks or crosses portages from one river to another, feeling for the time the joys of independence
and savage life.

The gaudy flies known as the Irish lake-flies, dressed on a small salmon-hook of about No. 1½, are successful throughout the entire length of the lake; but in the rivers a common brown or red hackle on the same sized hook, dressed with silver tinsel, scarlet body, and very full, long hackle, is decidedly the most killing, and in the lake answers full as well as the more expensive articles. Very small flies are not desirable, owing probably to the depth and occasional turbulence of the water in the lake and its discoloration in the rivers, which prevent their being perceived by the fish. Stout tackle and a heavy rod are better than lighter gear, as no one wishes to waste time on small fish, and the rises are so frequent that the angler will not become weary by continued casting. A gaff is necessary for the Mackinaw salmon, and a large landing-net for trout, but otherwise nothing is required different from that which the sportsman would take in a day's trip to the classic haunts of Long Island.

As the region around Lake Superior is well towards the Arctic zone, the weather is cool, and blankets, overcoats, and warm clothes are necessary; but there will be frequently several successive days of extreme heat, when the thermometer will rise to ninety in the shade. The great drawback to this section of country, in fact to all our unopened lands, is the immense number of mosquitoes, black-flies, and sand-flies. These pests are found numerously everywhere in our woods, but nowhere are they so plenty or combined so equally as along the shores of Lake Superior. All day long the black-flies watch their chance to find a bare spot of human flesh to sting and tear; immediately on the falling of the shades of evening the almost invisible sand-flies, the "no see 'ems" of the half-educated Indian, make their appearance in countless millions of infinitesimal torture, and all night long the ceaseless hum of the hungry mosquito drives sleep from the wearied sportsman's eyelids. Veils and ointments are, therefore, a prime necessity, without which a visit to this section is an impossibility; and even with the best protections, the warm days that give these insects unaccustomed activity are scarcely tolerable. But in spite of these petty discomforts it is a noble lake, beautiful in all its moods, silent and waveless in the warm sunshine, rippled and sparkling in the gentle breeze, or lashed to anger by the storm, when it rages along the shore and bursts in furious surf upon the rocks. Nowhere else can trout-fishing be had in greater perfection and more endless variety, nowhere else can the fisherman find purer sources of enjoyment or finer opportunities to exercise his art, and nowhere else can the lover of nature discover more to amuse or instruct him. It lies in the heart of an almost unbroken wilderness, the largest lake in the world, one huge spring of the coldest ice-water, and filled with trout that the painter can scarce find colors to imitate, and that will dwell in the angler's memory for ever.

6*

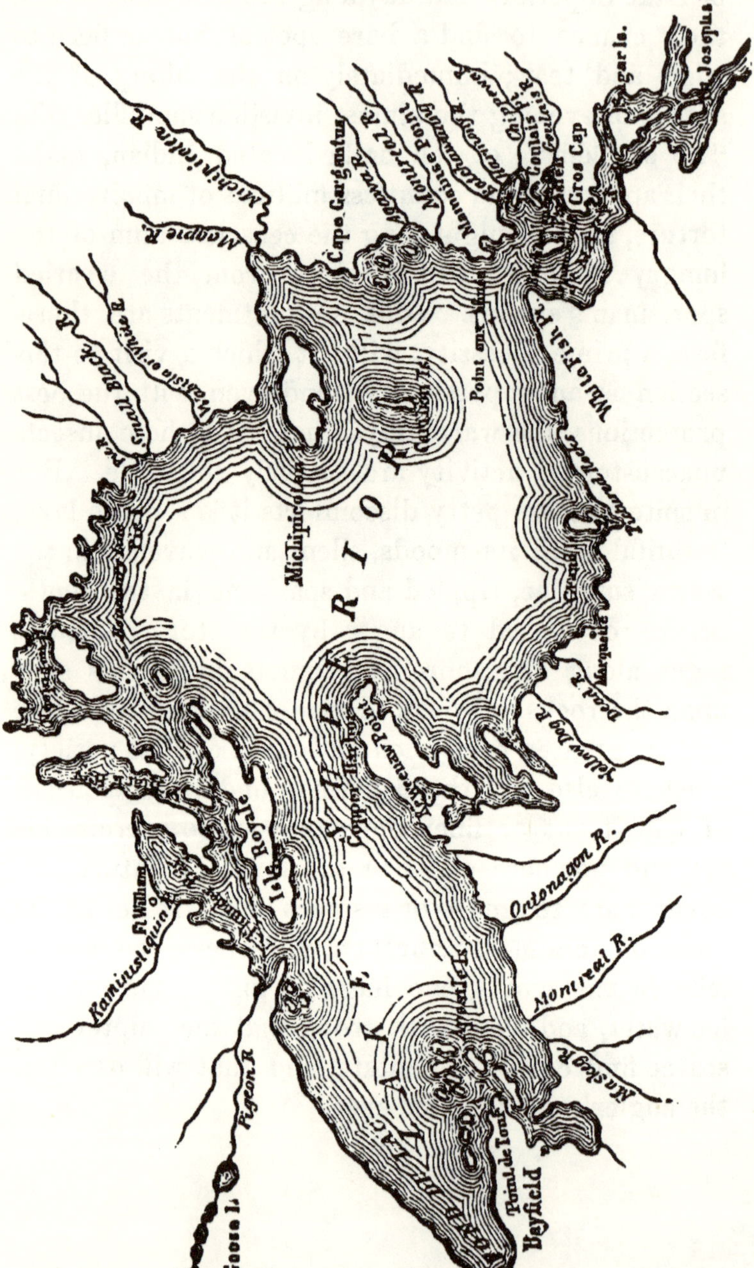
LAKE SUPERIOR
Michipicoten R.
Magpie R.
Pic R.
Small Black R.
Cape Gargantua
Montreal R.
Montreal Point
Gros Cap
Point aux Mines
Michipicoten I.
White Fish Pt.
Grand Is.
Marquette
Dead R.
Yellow Dog R.
Copper Harbor
Keweenaw Point
Isle Royale
Ontonagon R.
Montreal R.
Ft. William
Kaministiquia R.
Pigeon R.
Goose L.
Bayfield
Point de Tour

MACKINAW SALMON.

Namaycush—Salmo Amethystus.

OF all the varieties of *Salmonidæ* that permanently inhabit the fresh water, this fish, although
utterly destitute of game qualities, is alone entitled,
on account of his great size and excellence upon the
table, to the honored name of Salmon, is found
throughout the northern lakes, being prevented by
the impassable barrier of Niagara Falls from descending to the sea, occasionally visits Lake Erie, but
attains his finest condition around the cold, clear
depths of Lakes Huron and Superior. He is named
after one of his favorite localities, and reaches the
immense weight of nearly or quite one hundred
pounds, and is the grandest prize of the inland
waters of our northern continent.

In color, the Mackinaw Salmon differs, as does the
brook trout, according to the peculiarities of his
habitat, whether rocky or muddy shoals, or deep
open water ; and to such a degree that, according to
Professor Agassiz, he is known to the Canadian
Voyageurs under different names, and individual
specimens are frequently considered half-breeds or
a cross between this species and the Siskawitz.
Among the aborigines he is distinguished by the ap-

pellation which is usually spelled namaycush, although
it is pronounced namaegoose, and has the accent
strongly on the second syllable, and is never by them
confounded with any other variety of lake trout.
The fish of Lake Superior are of stronger colors; are
darker on the back; have redder flesh, and are universally preferred gastronomically to those of other
localities.

In Spring and early Summer, they appear to leave
the deep water, and seeking the rocky shallows, feed
voraciously upon the numerous small fry furnished '
in abundance by our western lakes. Throughout
May, June, July, and August, they can be captured
in abundance with the trolling spoon, trailed after
a boat propelled by oars or a gentle breeze, but are
rarely taken of over twelve pounds weight. At such
times they are excellent eating; their flesh being
rich, firm, and closely approaching in color that of
their congener, the famous *Salmo Salar*, and they
are delicious simply boiled or made into the basis
of a chowder.

Unfortunately, although they bite voraciously,
they give no play whatever, allowing themselves to
be drawn in without resistance, and there is no fish
approaching them in size which is so utterly devoid
of game qualities. At times they seem even to swim
gently forward as though they preferred coming
towards the boat, till the fisherman is uncertain
whether they are still on; and although at the last
moment they make a few flounces, their apparent
weakness for a fish so powerfully formed, is astonish-

ing. To be sure if a man had a hook in his mouth
he would follow the slightest pull; but we do not
expect such conduct from a fish, especially from one
endowed with the graceful and vigorous shape of
the Mackinaw Salmon.

They take any of the trolling spoons, appearing,
however, to prefer the old style, copied from the
bowl of a spoon, but rather elongated, to the expensive and fanciful modern improvements. Those sold
at the Sault St. Marie are from five to six inches
long and made of tin; but a better bait will be found
in the mother of-pearl imitation fish. To insure success, the weather should be moderate, either calm or
with a gentle breeze rippling the surface of the water,
for the reason that in the open lake a strong wind
will cause so heavy a swell that the fish cannot see
the bait, and the oarsmen cannot control the boat.
They are not shy; but as the water is frequently
deep, although wonderfully clear, the difficulty is to
attract their attention. For this purpose sufficient
line must be used to sink the bait slightly beneath
the surface, and the boat must not move too rapidly.

They are captured in all the bays and indenta-
tions of Lakes Huron and Superior, where the bottom is rocky and the water not over one hundred
feet deep. In Lake Superior they are abundant; in
Goulais' Bay, at Michipicotten Island, in the vicinity
of Bayfield, and almost everywhere else.

Late in the fall they retire to the sombre depths,
and are only taken by still fishing with a long line
and live bait, and at such times the deep water

abreast of Gros Cap is one of their favorite localities, and they are there frequently caught by the
Indians of from fifty to seventy-five pounds. They
are salted and smoked by the inhabitants for winter
use, but like the speckled trout are too dry for that
purpose, and should never be killed by the sportsman except as an article of immediate consumption.
They are usually distinguished among Americans
as the Mackinaw Salmon, although that universal and
totally undescriptive name Lake trout is occasionally applied to them, and are called by the Canadian
voyagers *truites du lac.*

The gums of this fish are of a purple tinge, and
from this peculiarity, which is by no means invariable, is derived their scientific name. The scales are
small and the lateral line is nearly straight. The
under gill cover is large and grooved ; while there
are many teeth, the prominent ones being very sharp
and much curved, and the tongue has a row on each
side.

The fin rays are :—D. 14, P. 15, V. 9, A. 12, C.19 $\frac{8}{6}$.
The tail is narrow at the root, and spreads broad
toward the extremity. The color on the back is
deep sea green, spotted with green and yellow spots;
on the sides it is purple, with lilac spots, and on the
belly pure white. The tail is dark and beautifully spotted the whole length. It is, altogether, a
remarkably handsome and graceful fish.

The spawning season is October, and the operation is performed in the shallows near shore, at which
time the fish are mercilessly speared by the natives.

LAKE TROUT.

Salmon Trout—Salmo Confinis.

THIS variety of the non-migratory *Salmonidœ*, although somewhat similar in general appearance to the foregoing species, does not attain the same gigantic size. It is found numerously throughout the middle and Eastern States, as well as in the great Northern lakes, but bears a vastly inferior rank in the estimation both of the epicurean and the sportsman.

Its gastronomic appreciation, I believe, however, is much influenced by the period of the year in which it is taken. Early in the season it is rich, firm, and of fine flavor, the flesh being of a light orange, and breaking into beautiful flakes. At such times it is unquestionably excellent. In Summer it is admirable as the foundation for a chowder, having some of the peculiarities in a higher development of the cod; and serving as a pleasant change from the ordinary boil or fry of the common trout. It is also quite eatable if cut into steaks and broiled.

Its scientific description is as follows:—The scales are minutely striate; the lateral line is slightly curved near the head; the tongue has large teeth

along the central furrow ; there are many acute teeth
on the palatines and vomer ; the tail has a sinuous
margin ; the bases of the vertical fins are spotted,
and the flesh is coarse.

The fin rays are :—D. 14, P. 14, V. 9, A. 12, C. 21 ⅔.

In color it is blackish or bluish-black, with numerous pale spots.   It is taken with trolling tackle, but
rarely or never with the fly.   The spawning season
is October, when it seeks the shallow water for that
purpose.

THE SISKAWITZ.

Salmo Siscowet.

THIS species has a dentition very similar to the *Salmo Amethystus*, but not quite so robust. The upper and lower maxillaries and intermaxillaries, and each of the palatines, have a row of teeth. The vomer one and the tongue two rows, beside the acute teeth. The tail is less furcate, and the dorsal fin is larger than in the Mackinaw Salmon. The flesh is rich and of fine flavor, but almost too fat.

The fin rays are:—D. 12, P. 14, A. 12, 14, V. 9, C. 30.

This fish is shorter and stouter, and not so distinctly spotted as the Mackinaw Salmon; it is altogether less handsome, but has similar habits, and bites readily at the trolling spoon. It was first described by Professor Agassiz, not many years ago, during his tour of Lake Superior, but has always been distinguished by the Indians and Voyageurs, and known among them under its distinctive appellation.

The Siskawitz inhabits the upper portion of Lake Superior, and never descends towards the outlet, and is taken in the neighborhood of Isle Royale in abundance. It is said also to be found in some of our other lakes, but is very rare.

Rock-fish—Librax Lineatus.

THESE glorious fish, the delight of the angler's heart, the bravest and strongest except the salmon, the largest without exception of the finny tribe that the sportsman pursues, frequent every cove and bay of our northern Atlantic coast, and furnish the main attraction of salt-water fishing.

Their mode of capture differs according to the locality; from the rock-bound coast of the Eastern States the adventurous angler, perched upon some projecting rock, casts the simple bait into the crested wave, amid the thundering surf of the stormy sea; along the sandy shores and in the tranquil inlets of the Middle States, gut snells, sinker and float come into play in the rapid tide ways; and among the numerous lagoons and bays of the Southern States the clumsy but effective hand-line is employed.

To the eastward, menhaden and lobster are the favorite baits; in Pennsylvania and New York shrimp, crab, and squid; and in the Southern States killeys, herrings, and other small fish. The artificial baits are the eel-skin, imitation squid, and gaudy bass-fly. The eel-skin used mainly along New England shores is attached to a hand-line, and cast into and drawn rapidly through the boiling surf of the ocean; the squid is towed with trolling tackle behind the sail or row boat, in the quiet waters of the Middle States; while the fly is used with stout rod and long line wherever the fresh current of some river haunted by fish falls directly into the salt water of the sea.

For casting with the menhaden from the rocks, New London harbor, Point Judith, West Island near Newport, Montauk Point, and Newport Island itself, are favorite localities; while the Little Falls of the Potomac at the Chain Bridge, near Washington, where the green waters dash over the sunken rocks and eddy round the cliffs that rise perpendicular from the river's brink, furnish the finest fly fishing for bass in the world.

For bait-casting the necessary implements are a large reel, running on steel pivots, two hundred yards of flax line attached to a 7° hook with a round head, and a rod of not over nine feet in length, with a large agate funnel top. With such tools experienced fishermen can cast a slice cut from the side of a menhaden, and weighing about three ounces, two hundred, aye, nearly three hundred feet into the curling breakers of the Atlantic ocean, and kill bass that will pull down the scales at fifty, sixty, and seventy pounds.

A mode of preparing a bass line to render it light and water-proof, without weakening it, is recommended by excellent authority, and is simply to soak it for one night in fish oil which does not rot linen, to hang it up to drain the following day, and to place it in mahogany sawdust to dry. When thus prepared it does not soak water, nor even sink.

Fly-fishing for bass, however, is the perfection of the sport, and infinitely surpasses in excitement all other modes of killing these noble fish. The best season on the Potomac is in July or August, and the favorite hours the early morning or the twilight of the evening. The ignorant and debased natives who inhabit the romantic region of hill and valley in the neighborhood of Tenally Town, about five miles northwest of Washington, and who, dead to the beauties that nature has lavished around them, and utterly unacquainted with scientific angling, look merely to their two cents per pound for striped bass, manufacture a fly by winding red or yellow flannel round the shank of a large hook, adding sometimes a few white feathers. They substitute for rod a young cedar sapling, denuded of bark and seasoned by age, and attaching to the upper end a stout cord, fish with the large flannel swathed hook in the rapids and below the falls of the Potomac, at the old chain bridge, and without a reel, kill bass of twenty or thirty pounds.

No spot can be imagined more wild and romantic, and with proper tackle, the reel, the lithe salmon rod, and the artistic fly—no sport can be more exciting. The roar of the angry flood, the bare precipices topped with foliage on the opposite bank, the flat dry bed of the stream where it flows during the heavy freshets, but at other seasons a mass of bare jagged rocks, and the dashing spray of the broken current lend a charm to the scene. While the fish, rendered doubly powerful by the force of the stream, and aided by the numerous rocks and falls, have every chance to escape.

The bass pursue the silvery herring, which is the principal natural bait, and ascend the Little Falls of the Potomac during the summer months in vast numbers. They are captured in such quantities with the net in the salt water and with hook and line in the rapids, as to be almost a drug in the market.

As the season advances, the native crawls upon some rock that reaches out into the stream, and with his coarse but elastic cedar pole, casts the roll of flannel, wrapped round a hook and misnamed a fly, into the seething current; and when the brave fish seizes the clumsy allurement the fisherman contends for the mastery as best he may, occasionally at the risk of a ducking in the stream consequent upon the sudden breaking of his tackle, and accompanied with considerable risk. When a man has but a slight foothold upon the slippery surface of a shelving ledge, and has attached to the end of his rod a vigorous fish of twenty pounds, he is apt to fall if the line parts unexpectedly. Many are the tales of such accidents, and now and then of fatal results. But with proper tackle, the scientific angler is master of the situation; he can reach any part of the current, casting into the eddies at the base of the precipitous cliffs opposite; he can yield to the rush of the prey; can retire, paying out line, to surer footing, and can follow the fish along the shore; and finally, having subdued his spirit and broken his strength, can lead the prize, gleaming through the transparent water with the sun's rays reflected in rainbow colors from his scales, into some quiet nook where he can gaff him with safety. Such is fly-fishing for striped bass amid the most lovely scenery, gorgeous in its summer dress of green and alternating hill and valley, dotted with pretty farms and smiling grain-fields; and there is but little sport that can surpass it.

Bass are also taken at the Grand Falls, ten miles further up the river; but the Little Falls are their favorite locality, as they are here just passing from the salt tide into the pure, sparkling, broken fresh-water. They frequently weigh twenty pounds, and occasionally much more; but, of course, the main run is smaller, and the number killed in lucky days is prodigious, being counted by hundreds.

Bass are said to be taken with the fly in other rivers of the Southern States, and also to a certain degree in those of the north. At the mouths of narrow inlets, where the tide is rapid and diluted with fresh-water, a gaudy red and white fly with a full body, kept on the surface by the force of the current and not cast as in fly-fishing, will occasionally beguile them; but generally speaking, bass are not fished for with the fly north of the Potomac.

Although the artistic angler naturally despises the miserable flannel abortion manufactured by the stupid boors of Tenally Town, it will often be found as

good a lure as though composed of the rarest materials; in fact the bass exhibit none of that daintiness
of choice that is universal with salmon. So long as
the fly is large and showy they seem to be satisfied,
and their immense mouths can readily grasp a No.
7 hook, such as the natives occasionally use. One
of half that size is abundantly large, however, and
the clearer the water the finer should be the tackle.
The rod, reel, and line are those appropriate to salmon fishing, although the line, if it is wet by salt-
water, should be afterwards rinsed in fresh to prevent rotting. Some fishermen fasten a float above
the fly, and paying out line let it run down stream
into distant eddies; but this is not so orthodox a
mode of proceeding, and does not require equal skill
nor as delicate tackle.

After a fish is struck, the same care has to be exercised if he is heavy that is necessary with the salmon, and he will often compel the angler to follow
him a long distance ere the gaff terminates the struggle. Bass make very determined but not such rapid
runs as their fellow-denizen of the flood, the *salmo
salar*, but rarely retain that reserved force which
makes his last dash so often fatal; nevertheless they
are resolute and powerful, and have to be handled
with care.

Another mode of taking bass, which is strongly
recommended, even for the open bays of the north,
by one of our best fishermen, but which I have only
tried in the narrow coves, inlets, and streams, where
the tide-way can be covered by a good cast, is to

use the salmon rod, line, and reel, but to substitute
a shrimp for the fly. The casting is then done in
the ordinary manner, and the gentleman referred to
claims, that it is by far the most killing mode. If
even equally successful, it is certainly far preferable
to the use of the float and sinker, or to the dull
monotony of bottom fishing. Any sport that brings
into active play the faculties of body or mind, and
which demands practice and experience, surpasses
the one that requires the merely passive quality of
patience.

The most successful, and excepting perhaps fly-
fishing, the most skilful method of taking the
striped beauties of the northern coasts, is with the
menhaden bait, cast into the boiling surf of the
ocean, or the larger bays; and this sport is universally enjoyed along the iron-bound shore of New
England, from New London to Eastport. This entire reach, is one mass of rock, indented by innumerable bays, or severed by inlets into barren islands,
where the tide rushes, and the surf beats; and in
every favorable locality are the bass taken with a
stout rod, a long line, and menhaden bait. From
almost every bold rock, or prominent island, can the
angler cast into the vexed water of some current,
made by the huge waves rushing over the uneven
bottom, and allure thence the fierce bass, who has
been attracted from the ocean depths, to feed on
the small fry that hide in the clefts and crevices;
and waiting with fins often visible above the tide,
to pounce upon his prey, mistakes for it the angler's

bait, and after a brave struggle surrenders to human ingenuity.

Although the true fisherman may pursue the small fish of the Delaware or Hudson, of New York Bay or the Sound, may patiently bide their time at Hackensac or Pelham bridges, McComb's dam or the hedges; and may have true pleasure in capturing them with dancing float and shrimp, or running sinker, and shedder crab; if he can spare a week or two, he should cut adrift from the noise and turmoil, foul stenches, and fouler deeds of the city; and hastening to Newport or Point Judith, enjoy the noblest sport of the salt water—bass-fishing with menhaden bait. He will need stout nerves, strong muscles, good tackle, and abundant skill; for he will be called upon to cast with the utmost of his power, perhaps a hundred yards, and to strike and land fish that may weigh half a hundred pounds. He will be exposed to the sea-breeze, or it may be the storm wind at early day-light, and the spray from the salt waves, and wet and cold will be his portion; but he will forget these trivial evils, when he strikes the bass of forty, fifty, or sixty pounds, the fish that he has been living for, and when he lands him safely on the slippery rocks.

Fishermen of character have been known to assert, that they could cast with the rod, the ordinary menhaden bait, one hundred and twenty yards; and although from a high stand, with the aid of a strong wind, this is possible, the ordinary cast is not over half that distance, and to exceed one hundred when standing on a level with the water is rare indeed. In fact, seventy-five yards is a good cast, and no man need be ashamed who can put out his line fair and true that distance. Rather better can be done with the hand-line than with the rod, but with far greater fatigue, and a painful over-exertion of the muscles of the arm that is almost unendurable to one who has not steady practice. The length of cast is in a measure controlled by the direction and violence of the wind and the elevation of the stand above the water; in a contrary wind the best angler will find it difficult to reach seventy-five yards, while from a high rock, with a favorable wind, he will cover that distance with ease.

The use of the hand line is neither artistic nor adapted to gentlemen who fish for pleasure, although more killing probably than the rival method. For rod fishing, the best tackle and implements are necessary; the rod must be short and stout, the finest being made of cane at a fabulous expense; the reel should have steel pins or run on agate, be made large and perfectly true, and the line must be from two hundred to three hundred yards long. Cane rods are preferred on account of their lightness and elasticity, but they are at present almost unattainable at any price, and the ordinary ones will answer well, although after several hundred casts weight will be found to tell on unaccustomed muscles. The objection to jewelled reels is, that a fall or blow may render them useless, while they run but little smoother than those with steel pins. The reel and guides must be large to deliver the line freely, and if the line is seen to bag during the cast between the guides, it is a sure sign that they are too small. The line is of twisted grass or raw silk, which is the best but most expensive and delicate; of plaited silk, which is the strongest; or of linen, which is cheap and common, but as they are all easily rotted, is the one in general use. The grass line, if it overruns and whips against the bars of the reel, is sure to cut, but it delivers beautifully; the silk line soon becomes water-logged and sticky; and the linen one combines these defects with a faculty of swelling when wet peculiarly its own. A perfect bass-line is a desideratum not yet supplied. The American reels and cane rods are perfection, but the lines are a cause of reproach and vexation of spirit.

Casting the menhaden bait is similar to casting the float and sinker, only the power is enormously increased and deficiencies proportionally magnified. The line is wound up till the bait, if a single one, is almost two feet from the tip, the rod is extended behind the fisherman, who turns his body for the purpose, and then brought forward with a steady but vigorous swing that discharges it without a jerk, like an apple thrown from a stick by rustic youths. The reel is so far restrained by pressure of the thumb that it revolves no faster than the bait travels, but does not in the least detain it, and upon the accuracy of this manipulation mainly depends the result. If too much pressure is used, the line cannot escape rapidly enough and falls short; if too little, the reel overruns and entangles the line, stopping the cast ere half delivered with a jerk that threatens its destruction. The fisherman must be able to use either hand on the reel to rest his arms and to take advantage of the wind.

If he is an adept he will drive the greasy bait straight and true directly to the desired spot, and if the weather is favorable and the fates propitious, he will bring up some scaly monster of twenty-five or mayhap thirty pounds, who will start seaward with bait, and hook, and line, and only be persuaded, after many efforts and determined rushes, that it is in vain. The strong ocean breeze will play with his hair and the salt spume wet his cheek; the vessels, like floating marine monsters, will drift across the waste of waters before him, the seagulls will hover round uttering their harsh cry, and he will cast and cast till arms and legs are weary, and he may kill in a single day a thousand weight of fish. The fresh air will give such a tone to his system, and the exercise such strength to his muscles, and the excitement such vigor to his nerves, that he will hardly believe himself the same relaxed, despondent, listless individual that left the city a week previous.

The most famous localities for the sport are West Island and Point Judith; the former is reached by the way of New London, and the latter by the Connecticut shore line of railway to Kingston. West Island has lately been purchased by a club of gentlemen, but will not probably be reserved exclusively for their use, as the neighboring islands being free to all no special privileges could be secured. There is often great difficulty in obtaining bait, particularly during a storm, which is the time that it is most needed, as the fish bite best in rough weather, and on going from the cities it is well to pack a few hundred menhaden in a box with ice and sawdust, and thus insure a supply for some days ahead.

VAIL'S.

POINT JUDITH.

It is a long, weary, and dusty ride by the way of
the New Haven and Shore Line Railroads to Kingston; but if, at the end of the journey, a pretty little
widow, with hazel eyes, is found waiting to drive
over to the South Pier in the stage, and you are the
only other passenger, you will probably consider
yourself repaid for all annoyances.

It is seven miles from Kingston to the South
Pier, the driver may happen to be a little tight,
very sleepy, and wholly unobservant of what is
passing in the back of his vehicle. Moonlight is
either reflected with great brilliancy from hazel
eyes, or else hazel eyes originate a brilliancy akin
to moonlight, and certainly moonlight, hazel eyes,
white teeth, rosy lips, soft hands, and a slender
waist, are very bewitching in a close carriage of a
moonlight night, with a preoccupied driver. Some
women have a smile like sunshine, and their laugh
rings like a chime of bells; and if you happen to be
riding alone with a pretty widow, and something
suggests love-making, and her merry laughter slowly
dies away into a gentle smile, and the smile fades
into a look of sympathetic feeling, that you have to
draw very near to see, till you feel her palpitating
breath upon your cheek, and her hand trembles

when by the merest accident you touch it, and the ride occupies an hour or more, you may, before the South Pier is reached, almost forget that you are married.

If this fortune befalls you at the station, you will probably fail to notice the beauty of Kingston village and Peace Dale as you pass through them, and will find the subsequent lonely ride from South Pier to Point Judith dull and dreary. Some two miles from the Pier is a house kept by John Anthony, the son of Peleg, where sportsmen most do congregate, and where all their reasonable wants, except the wherewithal to quench their thirst, can be supplied, and which is situated within a few steps of the best fishing stations. John Anthony is a Yankee born and bred, honest, faithful, willing, and acquainted with all the habits, devices, and iniquities of bass and blue fish. He will tell you that in May, when the grass plover have their long note, and are heard far up in the air travelling northward, bass are to be caught with the eel-skin; that in June, when high blackberries are in bloom, they begin to take lobster bait; but from July 1st, and all through the fall, they take menhaden, otherwise called bony fish or moss-bunker, the bait that the true and skilful sportsman loves to cast.

In July and August, the largest fish, occasionally bass of fifty and even sixty pounds, rejoice the heart of the angler by surrendering to his skill, while in the Fall, although more numerous, they are smaller. In both these particulars, the fishing at Point Judith

and West Island, and further northward, differs
from that in the vicinity of New York. Great suc·
cess, however, depends upon several contingencies.
It is supposed that the Gulf Stream, that prolonged
current of the Mississippi River, which sweeps with
its warmer temperature through mid ocean carrying
a genial atmosphere and fertilizing showers to the
otherwise arid shores of France and England,
changes its course yearly, approaching our coast and
sending its swarms of living creatures among the
rocks of Narragansett Bay, or withdrawing so as to
leave us desolate and to increase the severity of our
winters. We all know that our cold seasons differ
greatly in intensity, and bass fishermen know that
success in fishing varies equally; but from what
cause these results flow, no one can positively say.

After a heavy storm has darkened the water by
washing impurities from the shore, and at spots
where the dashing breakers fill the sea with foam,
the bass bite most fearlessly. Every crested wave
rising against the horizon ere it breaks, flashes with
their sparkling scales, and so sure as the bait cast
from the powerful two-handed rod reaches that
wave, so sure is it to be grasped by the nearest bass.
The breakers drive the spearing and other small fry
from their hiding-places among the rocks; the discolored water blinds them to their danger, and bass
trusting themselves in the very curl of the heaving
swell collect in myriads to the welcome banquet.
But as the discoloration misleads the spearing so it
also conceals from the bass the line attached to

the treacherous bait, and the latter, while pursuing remorselessly his prey, becomes himself a victim.

Neither shrimp nor soft crabs are used in this style of fishing, and the earliest bait, the eel-skin, is prepared by stripping the skin off the tail of an eel from the vent aft to the length of about a foot, leaving it inside out, and drawing it over a couple of hooks so placed on the line that one shall project near the upper and the other near the tail end. A sinker of the size of one's little finger is inserted at the head, and the bait is cast by hand and drawn rapidly. The rod is not often used in this style of fishing, as the heavy bait is apt to sink ere it can be reeled in. The skin is frequently salted to increase its firmness, and when used must be kept in continual motion, to the great fatigue of the enthusiastic angler.

The menhaden bait is prepared by scaling it and then cutting a slice on one side from near the head to the base of the tail, passing the hook through from the scaly side, and back through both edges, so that the shank is enveloped and the flesh is outwards, and then tying the bait firmly with a small piece of twine that is attached to the hook for that purpose. A menhaden or bony fish furnishes two baits, and the residue, except the back bone, tail, and head, is cut up fine, called chum, and thrown into the water to make a slick. A slick is the oil of the menhaden floating over the waves, and extended frequently by tide or current a long distance, attracts the bass, by suggesting to them that their prey is near at hand.

7*

Where the water is clear it is customary in rod-fishing, which is the only scientific mode, to use two hooks; the smaller, some two feet below the other is attached to a fine line or gut leader, and denominated without any apparent reason the fly-hook. Many of the best fishermen never use more than one bait, and where the fish are large and plenty, one is sufficient. The fly bait is not generally tied on, but twisted round the hook in a manner difficult to describe.

Lobster bait is deficient in tenacity, and has to be tied on like menhaden, and probably the natural squid would be an effective and manageable bait, could it be provided in sufficient quantities. Limerick hooks, except those manufactured expressly for the purpose with a round head, are in great disfavor, having a bad reputation for strength, and a stout but small cod hook is usually preferred. With skill, however, and plenty of line, the fisherman is more to blame than the steel, for the breaking of the latter. The best hook is now manufactured with a round head and is fastened to the line with two half hitches, the end again hitched above them so as to take the friction; and as it is carried off by the first blue-fish, or in the Yankee vernacular horse mackerel, that takes a fancy to it, the angler must be well supplied.

The Bait, especially a single one, is light, but experienced hands claim to be able to cast it more than a hundred yards, a feat that the tyro will scarcely credit; but ordinarily half that distance is all that is requisite. The line should not be less than six hundred and may be a thousand feet long, and if of flax should not be over fifteen strands. The rod, reel, and line, must be of the very best, and the guides and funnel top large, or the angler will fail to do himself justice, and will probably lose his largest fish.

The friction is so great in casting, that the thumb must be protected by a thumb-stall or cot, as the natives call it, or better yet, one for each thumb, so that you can cast from either side, and snub the fish with either hand. They are made of chamois leather, India-rubber, or some equivalent material; and in casting by hand, a similar protection is required for the forefinger. A shoemaker's knife is admirably adapted to cutting bait.

If, then, familiar with these things, you shall have chosen a favorable time during or at the close of a south-easterly storm, and at break of day, accompanied by John Anthony, shall have posted yourself upon Bog rock, or the Quohog, which is New England and Indian for hard clam, or upon the famous Scarborough, that great station in a heavy northeaster, you may anticipate brave sport. The waves will come rolling in, streaming out in the wind like a courser's mane, with snowy crest, and breaking with thundering roar they will sink back seething with foam. As the tide rises a few drops will fall pattering upon your feet; shortly the waves will leap up to your knees, then plunge into your pockets, reach to your waist, pour down your neck, and if you are not on the watch will lift you in their embrace and fling you torn and wounded down among the sharp-pointed rocks. You must wear water-proof clothes, and while you keep your eye on the line you must not neglect the inrolling swell, but avoid or brace yourself to meet its shock. And when the bass seizes your bait, and you have fixed the hook by one sharp blow, you must be gentle and moderate, only using severe measures where they are absolutely necessary. If the blue-fish comes, and he does not carry away your hook at the first snatch, reel him in as quickly as his indomitable pluck and vigor will permit. He is not game when you are bass-fishing. If the ungainly flounder, exhibiting unexpected activity, shall chase and grasp your bait, lug him out by main force, treating him, though excellent to eat, like the vulgar commoner he is.

When the day is advanced, and the game has grown wary, you may rest; and looking out to sea, perchance behold the blue-fish chase the menhaden and the porpoise devour the blue-fish, and the thresher shark plough his way through schools of lesser creatures, killing with blows of his powerful tail, and then devouring his prey at his leisure. You may listen to the " wild waves singing," and watch the continual change of the sky and water, enjoying the refreshing breeze and pure air, or amuse yourself by throwing in the head of a menhaden, and noting how quickly the bass that refuse your bait will strike with a great whirl at the floating object.

Two fishermen engaged with their sport were once standing upon a rock together, when one struck a very large fish supposed to weigh over seventy pounds. The sea was high and wild, and made it difficult to gaff the fish, after a wearying struggle had reduced him to submission. A favorable opportunity was watched when three heavy rollers had passed, covering the rock with spray, and the other fisherman darted to the edge of the surf to make the attempt. Unfortunately the bass, not being quite exhausted, made a short run that delayed the operation, till a gigantic wave, rolling in unheeded, caught the preoccupied fishermen unawares, engulfed them in its green waters, flung one down bruised and sore, and carried off the other who held the gaff, and was nearer the brink, into the deep water beyond. Poor fellow, he could not swim, and the terror of approaching death passed across his features as he looked up beseechingly and tried to cling to the steep and slippery rocks. The waves tossed him about like a plaything, bringing him close to the rocks, dragging him away, and then cruelly hurling him against them. His friend was powerless to save him ; but having a stout line, and the fish now floating exhausted upon the surface, shouted to the drowning man to catch the line and support himself by it. This was accomplished, and amid the dashing surf, alone with the shadow of death upon the water, the skilful fisherman, working his way carefully among the rocks, giving to the strain of the surging sea, but gaining every inch of

line the strength of his tackle would permit, led the
man and the fish, floating side by side, into a cove
that was in a measure sheltered from the fury of the
waves.

Slowly the line came in; the man lived, and still
clung to it, and although occasionally submerged,
managed to sustain himself sufficiently. Nearer and
nearer he came, quite close even to the shelving
rocks, and twice during a lull could have climbed
them in safety, had not his strength been too greatly
exhausted. He made a feeble effort, still clinging,
however, to the line, but was carried back by the
receding current, and it became apparent his life
depended upon his friend's ability to help him.

This was no easy matter; the strain upon the line
was excessive, the rocks were wet and slippery, and
the sea frequently swept across with resistless force.
Shortening the line as much as possible, the friend
crept down towards the edge, and taking advantage
of the first lull, called to the drowning man to cling
fast with his hands for a moment, and rushed down
to seize him. The instant, however, the line was
relaxed, the water carried away its feeble victim,
who was quickly beyond reach. Ere he could be
brought back a tremendous wave, resolute to devour
its prey, came thundering in; it rose above points
that had projected many feet out of water, it dashed
in flying spray high up upon those that it could not
overwhelm, its crest gleamed and hissed, and with
one mad leap it sprang over the intervening ledges
and threw itself upon the fishermen with fearful

power. The one upon the rocks was beaten down, and only by falling in a crevice and holding fast with all his strength was saved from being carried off. When the wave passed he struggled to his feet and looked down into the deep water for his friend. The line was broken, and man and fish were swept away together.

Danger never deterred a sportsman, but rather seems to enhance his enjoyment; and there is just sufficient risk and enough cold water to make fishing from the rocks a pleasurable excitement. The fiercer the storm and the wilder the water the better the fishing, and the peril is more than counter-balanced by the sport. Occasionally, at these times, a fisherman will be lost, but more frequently he will capture the gigantic fish that has been the ambition of his life; and if he does perish it is in a good cause, and he has the sympathies of all his ardent brothers of the angle.

Bass, like other fish, do not feed in a thunder shower, but during the latter part of a north-easterly or south-easterly storm, and immediately after when the wind has hauled to the westward and made casting easier, they are taken in the greatest quantities. In fact it is hardly worth while to fish for them at any other time.

At Point Judith there are some bay snipe and plover after the fifteenth of August, and the quail shooting which begins on the twentieth of September is quite good. Blue-fish or horse-mackerel are not pursued for sport, but rather pursue the angler, taking off his hooks and cutting his line with their sharp teeth most unmercifully. In fact a story is told of one that deliberately bit through the line above a large bass that had been hooked, and apparently released him designedly, from fishy friendship.

That excellent but neglected fish the porgee, which the inhabitants call a scup, is plentiful, and also the tautog or black fish.; and the bergall, which they denominate chogset or cunner, a worthless fish, is so abundant as to try the fisherman's temper by continually devouring his baits.

When the sea has subsided and the fishing is over, and you have as many fish as you want nicely packed in ice, you will have to drive over to the depôt behind the laziest horse, unless Anthony buys a new one, that it was ever your misfortune to ride after. The boyish driver, however, enterprising like his father, will poke and whip and utter that peculiar word comprehensible only to horse-flesh, "tschk," and if the animal does not absolutely lie down in the ditch you will make the seven miles in about two hours and a half, and be thankful that you have done so well; having reached home, what stories you will tell of the large fish you captured and enormous ones you lost, of the dangers you ran and how beautifully you cast, and your friends that receive of the game will believe in you.

THE SÓUTH BAY.

ONE cloudless day in the fervid month of July, a handsome, bright-eyed youth of something over twenty summers, opened the gate of the little yard in front of Deacon Goodlow's house and strode with an elastic step towards the side door. He was evidently at home and felt no need of ceremony, for without pausing to knock he turned the knob and entered.

The deacon's house was one of those innumerable romantic little white cottages with wings added after the main structure, that dot the flat surface of Long Island, or Mattowacs, as the poetical Indians once elegantly named the wonderful sand-bar; it was hidden in trees and almost covered with vines, and had an air of superiority and taste somewhat unusual.

"Well, Katy," said Harry, addressing a sprightly, rosy-cheeked maiden that he encountered inside, busy at some pottering woman's work; "what do you think, now? Your father and mine are going fishing to-day. I left them talking it over, and arranging that they were to drive over in your father's buggy, as our solitary horse is needed for other purpose."

"I am glad of it, Harry; Mr. Hartley takes too

little recreation, and father does so like a day on the
Bay. He was speaking about it only yesterday."

"But how odd that they should go alone; I
wonder why your father does not take you, you like
the Bay almost as well as he does."

"Pretty nearly," she replied with a laugh; "I
love the breeze and the water, especially when we
run outside and plunge into the monstrous waves of
the ocean. It seems so fresh, and limitless, and
powerful."

"Yes, and you like to pull out the blue-fish; it is
not all poetry, for to tell the truth, I have always
felt convinced from your way of looking at them,
that every time you caught a fish you thought of
the pot and fancied how nice he would be on table."

"Take care, sir, or the next time we go I will
leave you to your own devices in the way of cooking. Do you remember when I found you trying
to cook a big blue-fish on a long stick, over a huge
hot fire, without any salt or butter?"

"But the old folks will be sure to fall out over
politics or polemics, and come home in a dudgeon,
as they have been near doing before this, your
father is so fiery; I hope, for my future peace, his
daughter does not take after him."

"Now, Harry!" accompanied with a deep blush,
was all the answer, and Katy was turning away,
knowing instinctively how to punish her saucy lover,
when Harry hastily continued:

"I think I have prevented that, however."

"Have you? How?"

"I suggested something else for them to talk about, that will occupy their thoughts most of the time."

With a shy, sidelong glance, like a bird alarmed but uncertain of the danger, Katy replied :

"And what subject was that, pray?"

"Our love, Katy."

"A very silly subject, that need occupy nobody any time at all. You had better say your love, sir."

"Now, darling, don't tease, I have only a moment, or I shall be too late for the cars."

"Then, why not go at once? I am full as busy. Was not that Jane calling me?" She made.a great show of leaving, but managed to remain, evidently anticipating something of importance from her lover's manner, and in a female way dreading though desiring the disclosure.

"Wait one instant; I need not repeat how I love you, you have heard that often."

"Yes, indeed."

"But to-day I am to be admitted to a partnership with my old employer, who kindly offered it, with some complimentary remarks, so late as yesterday."

"You deserved it long ago."

"Not at all, I was well paid for my services; but now"--having drawn the willing but skittish beauty towards him, he whispered—"now I can keep a wife."

Her lips were close, her cheeks were tempting, her eyes turned away, her hands busy with the but-

tons of his coat, it is not certain he took advantage
of these opportunities; but suddenly starting into
life, she gave him a gentle tap on the ear, pulled
away, and turning to hide her blushes, called out, as
she darted from the room:

"You must catch her first, and the train starts in
twenty minutes."

"So it does," he muttered, as the delighted look
of admiration with which he had regarded her
faded slowly from his eyes; "what a darling witch,
it is so full of fun, and yet, as the neighboring poor
can testify, so gentle, generous, and sympathetic."
A thousand thoughts of all the loving acts he would
do for her came into his mind as he hastened towards
the depot.

"Well, friend," said Mr. Hartley, as the two
deacons were journeying along at a sober gait in
the old-fashioned but comfortable buggy of the
wealthier, "what a beautiful day it is, not merely
for our sport, and it could hardly be better, but to
admire the beauties of nature! The summer foliage
looks truly gorgeous in the broad sunshine."

"Yes, indeed, and the influence of such a day
must be felt by the moral nature of man. Even upon
man debased by vice, I believe in the country as a
moral purifier, and think a system should be devised
by which criminals would be thrown in contact with
it as much as possible."

"I agree with you fully, and had an evidence this
morning how it opens the heart and emboldens the
affections. You know Harry has long been atten-

tive to your daughter Katy, and I believe they have had a sort of half understanding."

"A fine fellow is Harry; true, honorable, and energetic," said Mr. Goodlow, heartily.

"He is so, and I, as his father, am proud to admit it; but Katy is a noble girl, and worthy of the finest fellow in the world."

"Well, we start the subject with a hearty accord," replied the friend, smiling; "I can readily imagine what will follow, and have no doubt we will be equally of accord on that."

"The short of it is, Harry has just been placed in a position that authorizes him to marry, and he wants you to trust Katy to him. On the subject of support he was satisfactory, and on that of love enthusiastic. He hoped your favorite minister would perform the ceremony."

This last remark was uttered very slowly, for it must be known the two deacons belonged to rival churches and different persuasions, and had had many a contest over form and ritual.

"That is a matter of small moment," was the response, "but if any form should be simple it is the marriage ceremony. I really think it had better be performed in your church, where there is less regard for formality."

"And for that reason I coincided in my son's selection; our church teaches us that while we are not to insist upon forms as the essence of religion in any of its departments, we are not to indulge prejudice against them. That they are immaterial either way."

"A strange view, indeed," responded the opposing deacon, warming to the question; "strange that any one could conceive that the form in which he expressed his adoration was unimportant; in all religion, prayer takes the form of the bowed head and bended knee. Unseemly postures and acts are themselves irreverent, not to advert to the effect they must produce upon the mind that indulges in them on serious occasions. We owe to our fellow-men respectful deportment on solemn occasions, how much more so to our Creator. Form is the embodiment of the spirit of true worship, and partakes of its essence and beauty."

"We fear," responded his associate, "that form, from its very beauty, may distract the heart and engross the attention to the neglect of the essentials of devotion. Pleasing forms are beautiful to our senses, but God looks to the pure heart and humble mind; the formalities of religion too often hide an aching void of real principle, and while they quiet the conscience produce no good fruit in the soul. Therefore, we dread them, lest though the sepulchre be whited on the outside it hide rottenness within."

They were both intelligent men, devoted to their sects, which although in belief almost identical, in forms were dissimilar; and they enforced and illustrated their views with great vigor, learning, and eloquence, and with the ordinary effect of religious discussions, that each was finally more firmly convinced that he was in the right. The hopes of their children were forgotten for the time, an occasional sharp innuendo added spice if not acerbity to the argument, and before their destination was reached a feeling of coldness, approaching dissatisfaction, had sprung up between the two friends.

There were no blue-fish running, and it was determined to try the striped bass that, although small, had begun to be plentiful, and in case of their absence to tempt the flounders, sea bass, black fish, or other like plebeians. In silence they pulled off to the fishing ground, and silently they cast overboard the anchor-stone and baited their hooks. Fishing has a calm, soothing influence incompatible with anger or estrangement. Occasional remarks were made which would doubtless have soon led to a perfect reconciliation had not the Fates prominently interfered. Mr. Hartley, who rowed the boat, had stationed himself in the bow, and strange to say began to take fish as fast as he could land them, while Mr. Goodlow, in the stern, usually the favorite location, caught nothing.

Fishing is a contemplative amusement, but when one contemplates his associate catching all the fish the amusement vanishes. Deacon Goodlow was a devotee of the gentle art, fancied himself an expert, and never doubted his far excelling his less experienced brother; had great faith in skill as opposed to luck, having often expatiated upon the fact that he rarely found an equal, and felt fully convinced that in skill he was not excelled.

Now skill is a very necessary thing and will tell in the long run, but luck is sometimes, doubtless for a wise purpose, permitted to triumph over it. In vain did the unfortunate deacon renew his baits, change the depth of his sinker, fish on the bottom or near the top; the result was the same. His irritation increased and broke forth into ejaculations of impatience, and a sudden desire to move to some other spot.

"There seem to be no fish here, we had better try a new place," he said pettishly.

"I am doing very well, and doubt whether we could better ourselves," replied his associate with that hilarity that success engenders, landing two bright little bass at once.

"You do not call that good fishing, they are mere sprats. I have taken many a bass of twenty-four pounds, and two of over fifty."

"But you know the run is always small in this month."

"Of course I know that; but I never saw such luck, you must have taken twenty, such as they are."

"More than twenty, thirty at least; but perhaps we had better change places, I have taken more than I want and you had better try your hand."

After some demur and a coquettish but half sulky refusal to deprive him of his "good luck," Mr. Goodlow complied with his friend's suggestion, but wonderful to say the luck changed at the same time; the fish all fled to the stern of the boat and were landed there faster than they had been previously over the bow. In fact, one line seemed to be bewitched as though the fish were in a piscatorial conspiracy. Even when the unfortunate fisherman extended his line and allowed his float to swing round beyond the stern and even alongside of his companion's, that of the latter would be dragged under at every moment, while his would remain undisturbed.

"Well, I have seen luck before," he began, fiercely, "but never such luck as this; how deep are you fishing?"

This question, as betraying the possibility of inferior judgment, fairly stuck in his throat.

"About three feet."

"Mine is the same. No, it is mere luck, that is all." Anger was making his language slightly ungrammatical.

Mr. Hartley replied, as he landed another brace: "Of course it is, and now let's change seats again and see if we cannot outwit the fish."

Being patronized by an inferior fisherman is almost unbearable, it implies triumph with nothing to justify it; and an assumption of superiority will be suspected if not intended. So Mr. Goodlow held out for a time, saying slightingly: "Oh, it was a mere question of luck, mere luck that must soon change;" but as it did not, and as his friend's manner was soothing and even submissive, he at last consented, with the air of conferring a favor, to resume his old place in the stern.

At the first cast which Mr. Hartley made after returning to his seat at the bow, he hooked and landed the largest fish yet seen. This was too much,

8

and if people swear inwardly it is greatly to be feared the unfortunate deacon will have to report hereafter one of the commandments broken on that occasion.

"Come," he said, "we will go home; another time perhaps I can have a little luck. I used to think there was something like skill in fishing, but there does not appear to be in catching these miserable little fish."

"Why, my last one must have weighed two pounds."

"Two pounds! Not an ounce over one. I have had enough for this day, and the sun is remarkably hot."

"Oh, I cannot go just yet; here comes another, nearly as large as the last."

"I insist upon it," Mr. Goodlow continued, having reeled up his line and taken apart his rod. "I will not stay longer, my horse must be fed, and it is late."

"When a person comes out fishing," replied Deacon Hartley, growing irritated, "it is a poor way to be wanting to go home because another catches the fish, especially as I am perfectly willing to divide equally."

"What do you think I care for those puny little fish? You may keep them all, in welcome."

"I suppose I may if I wish; they are mine because I have caught them, or nearly all; but I will give you half if you will cease grumbling at what you call your luck."

" Well, what is it if not luck! Perhaps you think
you surpass me in skill and experience," answered
the other sneeringly. "I tell you I am going home.
It is my horse, and you may come or stay, as you
choose."

With that he seized the oars and shipping them
into the nearest rowlocks, commenced furiously
rowing the boat stern first. But the anchor-stone
was down, and although he dragged it a few inches,
he did so slowly and with great labor. Mr. Hartley
went on deliberately fishing, but of course could
catch nothing while the water was being disturbed.

" Pull up the anchor-stone, sir," said Mr. Goodlow
fiercely, the perspiration streaming down his face.

" I will do nothing of the kind," responded Mr.
Hartley.

The tugging at the oars was resumed, but when
Mr. Goodlow was nearly exhausted, whether by
accident or not will probably never be known, the
oar slipped along the surface throwing a shower of
water over the quondam friend, fairly taking away
his breath. Without a word the latter dropped his
rod, and seizing the bailing scoop, a sort of wooden
shovel with a short handle, dipped it full of water
and threw the contents in his companion's face ; the
latter replied with a fresh douche from the oar.

The water fairly flew in mimic cataracts for ten
minutes, till both parties were wet to the skin ; originally, scoop had the best of it, but as skin and
clothes will not take wetting beyond a certain degree, oars caught up, and the two irate lights of the

church were as well drenched as if they had fallen overboard. Mutual exhaustion produced a cessation of hostilities, and after a moment's pause, Deacon Hartley slowly drew up the anchor-stone, and Deacon Goodlow rowed silently to shore. Without a word, without a glance, the latter stepped to his buggy, untied the horse, jumped in and rode off.

Mr. Hartley had to secure the boat, collect his fish, unjoint his rod, and walk four miles home. The day was hot, the road was dusty, the fish were heavy, and tired enough he would have been, if an acquaintance passing in a wagon had not taken him up. The dust having covered him from head to foot helped disguise what had happened, and he allowed the gentleman to think he had slipped into the water.

The thoughts of the two deacons on the way home were not enviable. One had to meet a son, the other a daughter, and the latter dreaded the interview most; not that he admitted he was most to blame, but fearing more her sharp eyes and reproachful countenance.

"Oh, Harry," said the pretty little girl usually so gay, now with sad-looking tearworn eyes, as she encountered her astonished lover on his way home from the railroad, " your father and mine quarrelled dreadfully to-day, so much so that they would not ride home together."

" Just as I expected," replied Harry, triumphantly ; " your father is so easily excited."

" No, but he says it was your father's fault, at

least he does not say so directly, but what he does say gives me that impression. Just think, your father threw water over mine, and he was all mud and dirt when he reached home."

"Impossible," said Harry, with a laugh, "he must have fallen overboard."

"Oh, no, and your father would not ride home with him."

"How did he get home then? he certainly would not have walked by preference four miles, on so hot a day as this. Imagine his half killing himself to deprive a person of his company who wished to be rid of him."

"Oh, it must be; father was so angry, he told me I should not see you again."

This response was illogical, and went far to disprove itself, but was enforced by her bursting into tears. "I have been crying ever since," she sobbed.

Harry consoled her, sure of her affection; and knowing that parents are a slight affair against affection, he brought back smiles to her lips by his comments on her account of her father's statement, and promised her it would come right if she only kept on obeying as scrupulously as she was then doing. She punished him for this by flying away in her former merry manner, leaving him to seek an explanation at home.

"Father," he said, on arriving there and seeking him out, "how spruce you look; that is your best suit. Are you going to pay a visit?"

"I believe not, this evening; my other clothes

were soiled while we were fishing." Strictly true,
but not all the truth.

"The deacon across the way came home rather
muddy, they say. What luck did you have? Did
it rain while you were out? There was not a cloud
to be seen in New York."

The father felt it would be useless to evade the
question, and related the whole story, bearing kindly the good-natured comments of his son, between
whom and himself there was a feeling of friendship
as well as of affection.

"And now, father," Harry began, after the recital
was over, "and now how are you going to make up?
You will have to make the first step, because you
were not in the wrong."

"Or, more truly, because my son loves the
daughter of the person who has ill-used me. Are
you not angry at my being left to walk home this
hot day?"

"I should be, if that wagon had not come along;
everything depends on that wagon. You know it
was much pleasanter than riding with an angry
man."

"But then the dust; my clothes are ruined; a
new suit will diminish your patrimony, which is not
enormous."

"Then I'll make you a present of a splendid suit
of black on my wedding day. I am rich, at least in
expectation, being a partner and no longer a clerk."

"To tell the truth," continued the father, dropping the tone of badinage, "I did feel ashamed of

myself, and was arranging a little plan of reconciliation, when our servant girl brought word that Mr.
Goodlow had forbidden her drawing water from the
well."

Harry looked at his father with a surprised,
troubled, and slightly angry look. The well was on
Mr. Goodlow's land, but had been used from time
immemorial by both families, as there was none other
near. He began to think the matter was more serious than he had at first supposed.

"I felt this to be unchristian," continued his
father, "and could not bring myself to make the
first advance after it."

"I can hardly believe the story, and will cross-
question the girl," replied Harry.

It turned out to be true, however; the girl had
been going to the well, as Deacon Goodlow descended, "all mud," as she described it, from his buggy,
and he seeing her at first seemed inclined to avoid a
meeting, but suddenly changing his mind told her
angrily never to come there for water again. With
all due allowance for kitchen exaggeration, the fact
could scarcely be disputed, and Harry suddenly
burst forth:

"We will dig a well of our own; I have always
hated dependence for anything, even on her father,
and then we'll see—"

What they would see was not very clear, except
that they would see the well built, for Harry, with
his usual impetuosity, at once set about making the
necessary arrangements, his new position enabling

him to supply the requisite means. He engaged the men and selected the spot that very evening.

Next day the well was commenced and advanced rapidly towards completion, the water for family use being carted in the mean time from a distance in barrels. What the deacon over the way must have thought when he saw the excavation progressing and the water cart regularly every morning passing in front of his door, no one knows; for not a word did he say. He could not have had an easy conscience nor a pleasant time, however, for Harry had not put his foot on the premises, and consequently Katy's eyes were almost as full of water as the barrel.

It was a long way down to the region of water, and if truth, as is generally believed, lies so deep, there is no wonder it is rarely reached; but the effort was at length successful, and when the liquid vein was struck the crystal fluid proved plentiful, half filling the deep well.

The water carts ceased their journey, the workmen were discharged, Deacon Hartley had a well of his own, Harry felt independent; but there was something else wanted. The latter had not exactly evaded Katy, who he knew was pining to see him, but, feeling his pride hurt, had not taken as great pains as he might to have thrown himself accidentally in her way. She had felt this neglect, and now when his pride was satisfied hers was aroused, and she kept herself carefully in-doors.

It took a week to build the well, and a week had elapsed since—that was two weeks of misery, all because the fish did not bite as they should have done, and neglected scientific allurements for less artistic attractions. Deacon Goodlow was miserable, because Katy looked unhappy and reproachful, occasionally enforcing her reproaches with a sob or two. Deacon Hartley was miserable, partly because he was ashamed of himself and partly because it went against his whole nature to quarrel; Katy was miserable, because her lover had neglected her, and she had had no chance to disobey her father's injunctions not to see him; Harry was the most miserable of the party now that the excitement of achieving his independence was over, because he missed the presence of his lady-love, and knew in his heart he had vented a little of his anger by neglecting her.

Harry was pining for her now in a much more rampant way than she had previously pined for him, and had revolved twenty impracticable schemes of restoring matters to their condition previous to the war. The inevitable laws of nature, however, that had caused all these mental wounds, helped to bring them to a crisis and finally to effect a cure. It was Sunday morning, and Harry had resolved twenty times he would join Katy on her way to church, for she went before her father to teach a class of Sunday scholars, and twenty times resolved that he would not. His father had convinced himself as many times that neighborly ill-will should be corrected at a sacrifice even of a little pride, and as often that he could not make the first advance;

8*

when a small voice was heard at the door, and electrified them both. It was not a sweet voice nor the tone rich, in fact it might be called harsh and unrefined, but the sound was pleasanter to Harry's ears than any he had heard in two weeks. The voice belonged to the extra help of Mr. Goodlow's household.

"Please, sir, master said I mussent, but could we have a little water from your well?"

Harry and his father gazed at each other and then at the girl in wonder.

"Please, sir," she continued, seeing their bewildered air, and addressing herself to Harry in an injured tone, "our well has run itself dry. Ever since you built yours the water has been getting lower, and last night it all went. Master says it's on account of the elevation, but I say it's because yours is further down hill."

"Do you mean to say you have no water at all?" said Harry.

"But I do, then, unless you call mud water; we managed to make tea last night by tying a new bit on to the rope; but wasn't it bitter and gritty, though? You ought to have tasted it; but to-day it's as thick as paste, and you know we cannot send a water cart on Sunday."

"How did you manage for washing?"

"That's how it comes we have no water for breakfast. We had saved up a little that had settled the worst down to the bottom, but we did not have enough to wash, and Miss Katy, when she tried to

use the well water, came out all streaked, and used
up all that we had put by; because, as she said, she
would rather go without her breakfast than go dirty.
I guess I wouldn't, though."

"But why did you not send to us before?" said
Mr. Hartley, compassionately.

"Why, because master thought as he had ordered
away your girl, you would do the like by me; unless he begged pardon, or something of that sort,
and he did not feel equal to that after your throwing him overboard the day you went fishing."

"He surely never said I threw him overboard?"

"No, but I guessed it; how could he 'a got so
wet otherwise, and why was he so mad?"

"Well, you guessed all wrong; I did nothing of
the sort, and hope you have told no one such a silly
story."

"Never mind that now," interrupted Harry.
"Mr. Goodlow is waiting for his breakfast; so take
as much water as you want or you will be too late."

"Give my respects to Mr. Goodlow," added his
father, "and say he is welcome to water from our
well at any time, and that I regret it has injured
his."

"Yes, and you can add that father will call on
him this evening, and now be off; I'll draw the
water for you." This was very polite in Harry, but
respect for woman, even in the humblest ranks, is
ever the attribute of an American, and—it is possible Harry may have wished to send a message to
Katy. "Leastways," as the girl would have said,

Katy was hardly out of sight of her front gate when she heard a step she well knew.

"Oh, Harry," she said, turning a pair of sorrowful eyes upon him, that shot reproachful torments into his very heart. "How could you?"

The sentence was incomplete in its construction, but complete enough in its effects; it was enforced with a little sob and made Harry about as contemptible a wretch, in his own esteem, as if she had rehearsed a set speech of an hour's duration, depicting his enormities.

"I am so sorry, Katy. Do you forgive me, I have been wretched?" This was a good tack, and being borne out by his appearance and evident contrition, went a long way towards securing his pardon.

What exactly was said, the tones being low and the faces close together, will never be discovered, but light came back to Katy's eyes, color to her cheeks, and a smile, if nothing more, to her lips; and ere the church was reached a happier couple could not be found within it. Joy is doubly blessed if preceded by sorrow, and only those who have known its want can appreciate happiness.

That Sunday evening, as had been his custom, unbroken for many years till the last two weeks, Harry presented himself at Mr. Goodlow's gate and entered unannounced. It can hardly be said he was wholly undisturbed, but outwardly exhibited perfect composure, prepared to meet and determined to exhaust the worst. Courage dispels danger, and there was nothing and nobody to meet more terrible than Katy herself. She was in splendid spirits, full of fun, rendered more touching and gentle on account of the recent estrangement, and charmed Harry with the renewal of her former witchery. He gave himself up to the mere enjoyment of her presence, following her every motion with unwearying admiration, and never removing his eyes from her loved form. He seemed as though drinking through his eyes her graceful beauty, and experienced all those charming sensations that love alone bestows.

He had almost forgotten, basking in present joy and dreaming hazily of future happiness, there was an angry father in existence, when the latter gentleman appeared at the door. A gleam of surprise crossed his features, but Harry at once stepped forward and was in the act of boldly justifying his presence, when he saw another figure in the doorway—that of his own parent.

Mr. Goodlow slowly advanced, and extending his hand frankly to Harry, said :

" I am glad to see you, and hope you will forget the errors and weaknesses of humanity, and forgive me the annoyance my foolish and unworthy quarrel has caused."

" And you, Katy," said Mr. Hartley, " must do the like by me ; we have been guilty of wrong, and should only do worse by being ashamed to own it before our children, whom our example is most likely to affect."

Harry felt as though he had escaped from a build-

ing on fire, and at once recovering his elasticity, replied :

"No ; in quarrelling Katy and I never intend to
follow any one's example.   Do we, Katy ? "

" We only regret," she continued, evading his
gaze, "that a shadow should have come between
those we love so dearly."

" I hope, never to return," replied Mr. Goodlow,
" and that these weeks of folly and punishment may
not be lost upon us all ; but let us speak no more of
it."

. " We have something more serious still to mention," resumed Mr. Hartley, gaily.   " We have been
settling your wedding-day, and, Katy, you should be
very grateful, for I named an early one."   He took
her affectionately in his arms, for she had always
been like a daughter, and kissed her warmly while
she hid her blushing face.

" That is right, father," burst forth Harry, enthu-
siastically.   " I suppose you went on the principle,
' If 'tis well done, when 'tis done, 'twere well 'twere
done quickly.' "

" No, Harry, on an entirely different one," said
Mr. Goodlow, laughing heartily.   " On the principle,
that ' All's well that ends well.'   Though that is but
a dry joke, as far as we are concerned."

PROTECTION OF FISH.

THE subject of the protection of fish demands
the consideration of every political economist, as
well as of every sportsman in our country, or we
shall soon be reduced to the condition of France, and
forced to repopulate our deserted streams and lakes
and furnish to the people, with great labor and at
high price, one of their chief articles of food. In
olden times, during the epicurean days of Rome,
and later during the reign of the Catholic fast days,
the utmost attention was bestowed upon the preservation, protection, and improvement of fish; enormous revenues were invested in immense tanks
where they were fattened, and different species were
transported to countries where they were unknown,
and domesticated in unaccustomed waters. With
the advent of the Roman Catholic religion, several
foreign varieties were introduced into England,
among others the fat carp and the lean pickerel;
and fish ponds were invariably attached to monas-
teries and convents.

Although the religion that ordains fish-eating to
be fasting, having shrunk from its gigantic reach and
extent, is confined in our land to a small sect, and
the inhabitants of the waters are no longer a religious institution; fish must always constitute a con-

siderable portion of the diet of the poor, and an acceptable change, if not permanently agreeable, to the rich. Whatever serves for food to the people, above all to the lower class, deserves the attention of the statesman, and any practice that will tend to diminish its price demands the assistance of the philanthropist. Consider if the price of fish were suddenly to double, how far the injury would extend, and how much suffering would follow. When a gradual change takes place in the cost of any article of food, man adapts himself to altered circumstances, and the loss, though equally great, is not so perceptible as when the advance is sudden.

That the supply of this food can be exhausted, and its quality easily reduced, is painfully apparent; streams in the neighborhood of New York that formerly were alive with trout are now totally deserted. The Bronx, famous alike for its historical associations and its once excellent fishing, does not now seem to hold a solitary trout, or indeed fish of any kind. The shad that a few years ago swarmed up the Hudson River in numbers incomputable, have become scarce and quadrupled in price during the last decade. Salmon, most nutritious and noblest of fish, which in ancient days paid their yearly visits in vast numbers, if early historians are to be believed, to our principal rivers as far south as the Delaware, are at present taken nowhere to the southward of Maine, and in but limited quantities even in that wild region.

On every portion of our sea-coast, in spite of re-

plenishment from the mighty ocean, the same diminution is visible, while many of our confined inland
waters are absolutely depopulated. The insatiable
maw of New York market swallows alike the trout
from Maine, the bass from Lake Erie, or the white-
fish from the Sault Ste. Marie, while the parvenus
that have acquired sudden fortunes in that wonderful city, endowed with the instincts of neither gentlemen nor sportsmen, think it magnificent to devour
trout in Autumn and black bass in Spring, judging
by their extravagant price that they must be rare
and therefore good. The rapidity with which a
section of country can be fished out by energetic
pot-hunters where the law .places inadequate restraint, and often in spite of the law's restraint, has
been remarkably evidenced in the history of Sulli-
van County. When the Erie Railroad was still incomplete, and the tide of explorers had just commenced to penetrate beyond Goshen, and only occasional stragglers reached the land of promise and
performance beyond Monticello; the swamps were
alive with woodcock and the streams with trout.
But as the railroad advanced and gave improved
facility of travel, so-called sportsmen poured over
the country in myriads, following up every rivulet
and ranging every swamp, killing without mercy
thousands of trout and hundreds of birds, boasting
of their baskets crowded to overflowing, and count-
ing a day's sport by the hundred; till Bashe's Kill,
where the pearly-sided fish once dwelt abundantly,
was empty, and the broad Mongaup, the wild Calli-

coon, and even the joyous Beaver Kill, with its innumerable tributaries, were exhausted. The woodcock disappeared from the cold black mud of the
springy swamps, the trout no longer broke the surface of the noisy rills of that picturesque region,
and the hunters and fishermen turned their attention and carried their clumsy rods, bait-hooks,
cheap guns, and case-hardened consciences, elsewhere.

So it has been and will be everywhere, unless the
people and the real sportsmen take the matter in
hand ; the farmers, who are after all to be the salva-
tion of our institutions, lose by the destruction of
game one of the greatest attractions of their lands,
and are interested in preserving for themselves and
their city friends the wild dwellers in the lakes and
brooks from wanton and ruthless destruction. Law-
givers are concerned in the passage of proper laws on
account of public interest, and the increasing necessity of cheap food that a rapidly augmenting population engenders. Sportsmen have the greatest
stake, for if they would retain for their old age and
leave to their children the best preserver of health,
a love of field sports, they must protect game-birds
and fish. They should discourage, by their conversation and example, all infringement of the law or
any cruel or wasteful prosecution of what should be
sport. If they find a man who destroys, for the
purpose of destroying, they should not only shun
but expose him ; if they meet with a case of palpa-
ble infraction of the law, they should enforce punish-

ment; by these means, and the enactment of judicious statutes, the beautiful wild creatures that form so pleasant an addition to the charms of country life, may be preserved in undiminished numbers for all time.

The first necessity, however, is that proper and uniform enactments should be passed in every portion of our extensive nationality. If the close times differ in adjoining states, fish will be killed in one and sold in the other; it is useless to attempt to forbid the catching of trout in Maine, if they can be eaten in New York. Pinnated grouse, killed on the western prairies where they are fast being exterminated, are sold openly in New York markets in consequence of their omission from the game law, during the entire spring, until the heat of the weather prevents their transportation. Black bass are frequently exposed on the hucksters' stands heavy with spawn, and pike-perch are hardly regarded as desirable in any other condition.

The universal rule should be comprehensive and simple, as the habits of the fresh water fish are sufficiently well known; protection should be given during the spawning season, and for such a period before and after as to prevent the annihilation of those who have survived the numerous dangers that surround them, and are ready for the duties of parturition, and to allow them to recover from the exhaustion resulting from the operation.

No trout should be killed except from the first of March to the first of October; no lake trout except

from the first day of February to the first day of
November, and no black bass or mascallonge from
the first day of January to the first day of June.
These times may be restricted for certain localities
where greater protection is necessary, but should,
under no circumstances, be enlarged. Trout spawn
from the middle of October to the latter part of
November, and do not recover their condition till
the opening of Spring. Lake trout spawn about the
same time, and mascallonge and black bass in
March, April, or even as late as the early part of
May.

None of these fish should be taken in nets, nor by
spearing, and no fykes, seines, or gill-nets should be
used in the waters which they inhabit. Stringent
regulations to this effect are necessary, as it has
been the habit of the market fishermen of the
northern section of our country to use a net with
meshes small enough to catch yearling trout, and
which they frequently throw to one side and leave
to perish miserably. This net fishing is continued
all winter, so that not only are thousands of large
fish destroyed in the act of spawning, or just after
doing so, but millions of the young, the seed of the
harvest, are slain without profit, being left on the
ice to freeze.

Spearing is also terribly fatal. None can escape
the sharp eye of the spearsman, and although many
more are wounded than killed they rarely recover, for
their natural enemies, the eels, are ever on the alert
for such occurrences, and fastening themselves upon

the wounded spot suck out the little life that is left. There are many streams of New Jersey which, by persistent gigging, as it is called, have been divested of every swimming thing, so that they are absolutely uninhabited. Not only trout, but catfish, eels, and suckers, have met the same untimely fate, and now boys and men search vainly for their prey.

By fair fishing no stream or pond can be entirely exhausted; when trout have the privilege of biting or not, they will exhibit sufficient circumspection to perpetuate their species; but when they can be followed during the hours of darkness to their retreats, and exposed by the glare of the jack, are liable to death by the fatal spear, or in case they may be enveloped by the all-devouring net, they have no defence or escape, and must soon disappear entirely. Their numbers, instead of helping them or delaying the catastrophe, excite the cupidity of the poacher, and accelerate instead of deferring their destruction.

Interested parties in various sections of the country, endeavor to convince themselves and others that trout change their nature in these favored localities, and either spawn from time to time as fancy dictates, or postpone the performance till winter's frosts have driven profitable visitors to their city homes. The proprietors of the frontier taverns, where sportsmen congregate in search of finny prey, boldly assert that there are several kinds of brook trout, of which one variety spawns in September, another in October, and so on in such manner that it is always right and proper to fish for them. Naturalists have, as yet, failed to discover this peculiarity or describe these varieties; and although they know that individuals may differ casually or delay the act a few weeks, they recognise one well known spawning season. The *ova* of trout are largely developed in September, and, except in the colder latitudes and where they are extremely abundant, these fish should be exempt after the first of that month; but in October and November, pressing hunger should be the only excuse for killing them.

The laws, however, are not so much to blame as the neglect of their enforcement; perfect statutes will not answer if they are not carried out, and the first duty of sportsmen's clubs and of individual sportsmen, a duty to humanity, to themselves, and to their fellow creatures, is to enforce the game laws. By game laws are not meant those barbarous statutes of England that made it more criminal in a poor man to slay a hare than a human being—statutes that are deservedly odious to free men, and which by no possibility could be introduced into the New World; but provisions for the protection and preservation of the wild inhabitants of our woods and waters, a common heritage of beauty and sustenance, and the property of our citizens indiscriminately. These creatures are a considerable source of wealth, worthy the most careful attention; they breed and increase of themselves without care or expense; and constitute a large portion of the stock of our markets. It would be an interesting investi-

gation to ascertain how much money is paid yearly
in the City of New York for the wild deer and
game birds of the west, the sea fishes of our coast,
the finer varieties of our inland waters, and the salmon of Canada. The latter, alone, amounts to
hundreds of thousands of dollars, and is a severe
tax paid to a foreign country for the fatuity that
drove those noble fish from our own rivers.

This vast source of revenue will, however, disappear, unless precautions are taken to prevent the
untimely slaughter of these unprotected creatures.
If their periods of incubation are disregarded, their
nests and spawning-beds broken up, and themselves,
when engaged in the duties of maternity, disturbed
or slain, they will diminish rapidly till the forests
shall cease to be vocal with their harmony, and the
water animated with their gambols.

In England not only do game preserves produce
a good rent from enthusiastic sportsmen, but the
fisheries, particularly of salmon, are extremely valuable as commercial enterprises. At present, in our
our country, we only recognise the value of these
advantages by their loss. The Tay produces a
rental of $70,000 yearly for the salmon fisheries,
and so profitable have fishing rights become, that
several rivers that were once exhausted have been
restored, and now yield large revenues.

If we would have salmon at our own doors, we
also must restock the Hudson, the Connecticut, and
the numerous other rivers that were once frequented
by them. But the trout and the black bass are still

with us, and by decent care and treatment may be plenteous, for the pleasure and support of ourselves, our children, and our children's children. Considerable attention has been expended upon some of the ponds and streams on Long Island; and although the poacher makes occasional depredations, and lurking through the bushes plants his net, or with wriggling worm draws forth his unseasonable prey during the forbidden periods, the improvement already is remarkable. Ponds that were once empty of fish are made beautiful by the splashes of the playful trout, and streams that were deserted are replenished. Enforce the law thoroughly, and discontinue unreasonable slaughter, and fish, from their enormous fecundity, must increase immensely.

It is probable that the localities in the neighborhood of our large cities have passed their worst days, and that the beautiful lakes and rivers, ensconced in the wild woods and amid the green hills of our unopened country, are in the most danger. A cockney sportsman, by which we mean not a city sportsman, but him who, wherever born or bred, fishes only for quantity, and from a vain-glorious spirit of boastful rivalry, is, indeed, a ruthless thing; he spares neither fish, flesh, nor fowl, whether he can use them for food, or must leave them to putrify, and regardless of the means or implements he employs. This merciless biped invaded Moosehead lake one year, armed with fly and bait rod, and with two additional trolling rods projecting from each side of his boat as he moved from place to place, murdered thousands of glorious trout; supplying his own wants, the public table, and the hog-pen—for the latter was separated from his feeding place—till the pigs, disgusted at his brutality, were surfeited, and bushels of putrescent fish had to be buried or thrown into the lake. Others, almost as murderous, roam the north woods of the State of New York, and even penetrate as far as the unbroken shores of Lake Superior, threatening annihilation to our game of every kind. The man who kills an animal, bird, or fish, knowing that it must be left to spoil, justifies the charge of cruelty against our class, and deserves the scorn and condemnation of all right-thinking men.

Wanton injury to public property, in game, should be punished precisely as similar injury to public property in grounds or buildings, by incarcerating the offender in prison; for of the two, the latter is less injurious in its ultimate results. A building may be replaced, but who can restore life to the fish that bears a thousand undeveloped young in its bosom, or can give back to the starving fawn the mother that has been slain at its side? Mere pecuniary fines are an insufficient punishment; the poaching criminal is the poorest, as he is the meanest, of offenders, and laughs at any attempt to collect penalties that are not enforced by imprisonment; while the wealthy cockney is willing to run the risk of fine if he can, by taking the advantage of honest sportsmen, have the chance of boasting of his wonderful prowess and suc-

cess. A few months in jail would cure the reck-
lessness of the former and cool the ardor of the
latter.

A still more murderous proceeding, so infamous
that it is rare even with professional poachers, is to
cast poison into the water, thus slaying, by one fell
process, large and small, young and old. Condemnation of such a practice is unnecessary; and were
it otherwise, fit language could hardly be found to
depict its enormity.

By the introduction of unsuitable fish much injury
is occasioned, more frequently through ignorance
than wilfulness. Perch placed in a sluggish trout
pond, like many of those on Long Island, will
devour the young fry, and soon diminish the yield;
and pickerel, which are especial pets of our farmers,
although nearly worthless for food or sport, have
devastated some of the best ponds in the country.
The former are devotedly fond of minnows or small
fish of any kind, and such bold biters as to give rise,
in England, to the story of a country gentleman
who enticed an ardent angler to his house by stock-
ing one of his ponds with several dozen perch, all
but one of which the visitor captured on the day
after his arrival, before breakfast. The pickerel is
exceedingly voracious, and also right fond of his
smaller fellow fish for dinner.

To meet these cases the ponds must be drawn off,
as neither perch nor pickerel remain in running
water, and the waters must be re-stocked. In fact,
wherever, from any cause, the drain is greater than

the supply, the deficiency must be made good by artificial means.

By these means can the seductive little beauties, whether of the feathered, furred, or scaly tribe, that allure us to the great woods, the pleasant meadows, or the sparkling brooks, be preserved through endless time in undiminished abundance, furnishing the incentive that leads us away from our dull books · or wearying cares, the crowded streets, the congregations of eager men, the trials and excitements of business, to gentle communings with the hills and skies, to contemplative musings beneath the leafy forests, or by the noisy water-falls, strengthening our nerves, renewing our hold of life, and elevating our moral nature.

FLY-MAKING.

BEFORE making an artificial fly, it is essential to ascertain and select the best materials, and the necessary implements for the purpose. In the Game Fish of North America the author has explained the simplest and easiest mode of tying a fly, and if there be any person who has not read that work he should procure it at once. The instructions there contained must be first mastered before the following are attempted, lest discouragement should result; and no one that does not desire great accuracy and finish need waste the time and labor of understanding and executing the ensuing directions. There are a few persons who wish to tie a fly handsomely; this chapter is written for them. The fish probably care little whether the fly is made at Conroy's establishment, of the finest materials and from the most approved patterns, or by some unknown German wholesale dealer, of any chance feathers.

Remember, however, that he who strives not after perfection never attains mediocrity, and the improvement of himself is one half of the angler's pleasure. If we are content with an ungainly fly, we will be satisfied with inferiority of rod and tackle; and although the fish may not see the difference, the angler may become, from neglecting one point, slovenly in all. A well-made fly is a beautiful object, an ill-made one an eye-sore and annoyance; and it is a great satisfaction both to exhibit and examine a well-filled book of handsomely tied flies.

Nothing can be thoroughly done unless strict attention is given to minutiæ. The material must be selected and protected with the greatest care, the scissors and knife must be sharp, the spring pliers of suitable strength, and the nails of the workman must be long and his hands scrupulously clean. Hereafter the table-vice, the use of which was recommended in the Game Fish of North America, and which will be found both convenient and for extreme neatness necessary, will be dispensed with, and the hook held in the hand during the entire operation. This at first may appear awkward, require more time, and give an inferior result; but sad would be the case if the loss of a vice were to diminish a man's capabilities.

The selection of the hook depends mainly upon the fancy of the fisherman, and partly upon the locality of its destined use. If fish are scarce and shy, select one that will insure striking ; if they are abundant, but strong and vigorous, choose one that will hold. In trout-fishing there are two that bear the palm in striking, the sneck bent and the Kirby bent Limerick; in holding a fish after he is struck, my preference is for Warren's Lake-trout hook, which, however, does not make a handsome fly ; for salmon-fishing, the O'Shaunessey Limerick is the general favorite. The objection to the straight or hollow-pointed Limerick, is that it may be drawn over a flat surface without catching, while the point of the O'Shaunessey, by projecting, catches and penetrates.

Fish-hooks of the best quality of home manufacture, of all shapes and sizes, may be obtained at from twenty-five to seventy-five cents a hundred, and will be found equal if not superior to any English hook at double the price, or they can be manufactured of any shape desired.

So few persons make their own flies in this country that none of the tackle-makers sell the materials, and hence the amateur will have to collect the latter as opportunity offers. Gut, of course, can be purchased anywhere; but the strongest kind of that suitable for salmon-fishing is often difficult to obtain, if not entirely out of the market. In trout-fishing, select fine, round, transparent strands, and pay from one to two dollars per hank of one hundred strands; for salmon choose the strongest and roundest, and pay from three to four dollars. Gut is imported from Spain and Italy, and is made by drawing out a dead silk-worm till it is of the proper fineness; and none imported from the East, and no imitation of grass, sinew, or the like, is worth using. The quality can be determined by its hardness; if it resists the teeth well, it is good; age weakens and finally decays it.

The best wax, although it is by no means perfect, is made of one part of resin, one of beeswax, and four of shoemaker's wax, the two former melted together and poured into water, and then worked in

with the latter. It should be kept in a small piece
of leather. Shoemaker's wax itself is the strongest,
but is sticky in warm weather and hard in cold.
The best silk is the finest sewing-machine silk,
marked with three 0's on the spool; but for very
small trout-hooks the better plan is to twist two or
three strands of spool floss-silk together and wax
them carefully.

Tinsel of a superior kind is difficult to obtain; the
silver should be both variegated and plain, and the
yellow either gold or well covered with gilt, and
both flat and wound over fine silk. A mixture of
both sorts of a poor quality is used to tie linen
goods, and can be obtained at the furnishing stores,
but a better article is to be had from the importers
of gold and silver braids. The proper kind of floss-
silk comes in spools, and can be wound off by the
single thread over the hand till a proper thickness
is attained, and will work much better than the
common floss skeins. If the latter are used, they must
be divided into several strands and are apt to bunch.

Worsted of all colors can be obtained in the
rough, or the yarn may be picked or used intact;
the former is the best plan, and rivals mohair in
appearance.

Mohair may be purchased from the importers of
woollens, while it seems impossible, except by direct
importation from the English tackle-shops, to obtain
either pig's hair or seal's fur. For salmon-flies the
two last are infinitely preferable, having a gloss that
no other material possesses.

Mohair and camlets are the finest selection of goat's hair (the former being carded and the latter combed), and work beautifully. The most elegant flies are those with silk bodies, but they are rarely so effective as those of mohair. Many of the wild animals of our woods furnish a fine fur, such as the grey, red, and black squirrels, martin, mink, rabbit, and others.

A golden pheasant is indispensable for salmon-flies, and a spoiled skin can be obtained from the taxidermists at from two to five dollars, according to their scarcity. Hackles for salmon-flies should be large and from matured cocks, those for dyeing delicate colors pure white; while for trout-flies they should be small, either from hens or from cocks not over two years old, and taken from the upper part of the head. They must taper well to the point and not have a stiff stem, and should have the fibre about the length of the hook shank. For wing-flies they must be smaller than for hackle-flies and palmers, and the superfluous fibres are to be stripped off before the feather is tied on. Small neck feathers of almost any bird will make a hackle sufficiently large for the midge flies. The natural colors afford abundant variety for trout-flies, but for salmon the gayest must be dyed. The necessary colors are red, claret, blue, orange, purple, and yellow; and by suiting the dye to the natural color, so that the latter shall shine through, a fine effect is often produced. Considerable practice and experience will be necessary in selecting hackles to distinguish the weak from the harsh, and to determine the proper size and elasticity. Collect all varieties of dimension and color, and tying each selection round the roots with a thread, keep them in separate papers. After a while, those that experience shall have proved to be unsuitable may be discarded.

The feathers of small birds make good wings for trout flies, and there is not generally much difference in their color. Our brown thrush is nearly the shade of the English land-rail; the robin furnishes a fine and cohesive feather; the woodcock's tail makes a pretty fly, while the mallard and wood duck are indispensable.

There are two distinct feathers from the mallard which are used for different flies; the brown and grey mallard feather, both taken from the drake, the former from the back near the wings, and the latter from the body beneath the wings. The bird must be in good plumage, and under the most favorable circumstances they are both, except in simple wings as hereafter described, difficult feathers to tie; the fibres, although very fine, being apt to separate. Another light feather, much easier to handle than the grey mallard, is taken from the back of the canvas-back, but is of rather too pale a color; that from the red-head is of darker grey. For salmon flies a larger range is requisite. The turkey of all shades, but especially the black and brown of the wild bird, is the main-stay; the golden pheasant's tail is somewhat similar; the peacock gives us excellent feathers of many shades, and the finer herls from the eyes of

9*

the tail add lustre to a mixed wing. Peacock and ostrich herls are used for the heads and bodies of certain specimens. Ibis, macaw, guinea-fowl, blue-jay, king-fisher, parrot, are all necessary; while the Argus pheasant, although injured by the water, makes an exquisite wing, and the silver pheasant is used with effect in black bass flies.

For dyed feathers the pure white of the swan furnishes an excellent material, while crossing colors, such as yellow over ibis, produces great brilliancy. The mallard and canvas-back are also favorites for dyeing. The principal shades are yellow, blue, and purple.

We will now proceed to make a salmon-fly after the simplest plan on a large hook, and remember that the point is held down, and when the further side is spoken of, it refers to it in that position; the head is always towards the right and that is called the upper part, and towards it is above.

Select a piece of stout gut a little longer than the shank; pare down the ends with a knife; double them together so that one shall extend beyond the other; insert the picker between them, bend at the top and shape it by twisting and pinching the ends. If the hook is very large it is well to take several strands of gut and first twist them together by means of a vice fastened to each end, while they are wet and before shaping them over the picker. When the gut is prepared lay it down and take a well waxed piece of silk about six inches long, and holding the hook in the left hand, wind a number of

separated coils from the lower towards the upper
end of the shank, but not quite to the head. If the
silk is well waxed it will remain in its place while
you pick up the gut with your right hand, and lay
it along the under side of the shank upon these
coils. Hold it there with your left while you wind
firmly and closely toward the bend; catch the last
turn beneath the gut or pass a half hitch, and cut
off the end. Take a fresh piece of silk, always
thoroughly waxed, and pass a few turns over its end
so as to fasten it; then hold a piece of tinsel four
times as long as the shank between your left forefinger and the further side of the hook, just projecting above it, and nearly vertical; pass three turns
over it, and wind the silk in separated or loose coils
towards the head and let it hang there. Fasten the
spring pliers on to the lower end of the tinsel length-
ways with it, and holding the shank in the right
hand, with the left forefinger in the pliers, twist
several turns down and then back to form the tag,
covering the edges of the first turns with the second
carefully and neatly; let the pliers hang; pass the
hook to the left hand; unwind the silk with the
right down to the tinsel; fasten off with three turns
and cut the tinsel close to the hook. Unwind from
the floss-spool over your right hand a dozen strands,
and smoothing them evenly together and holding
them against the hook with the left, tie in the ends
firmly, and again coil the tying silk toward the head
out of the way. You may wind the floss with either
hand or with the pliers as you please; if you wind

with the right hand, hold the hook in the left and
press the second finger on each turn as it is passed;
this is called stopping it or using the stop. After
covering about one sixteenth of an inch, seize the
end between your second and third or third and
fourth fingers, and hold it firmly while you bring
down the tying silk and pass three turns; holding
the silk in that way is called using the catch, and is
difficult to acquire with facility. Cut the floss off
neatly, and selecting a feather from the golden
pheasant top-knot, lay it on its face,—the side of the
feather which lies nearest the bird from which it is
taken, is the inside or back, and the contrary side
the outside or face,—and secure it firmly. Stop the
tying silk and take up your hackle, which should
have been previously prepared by stroking back and
pulling out a few fibres toward the point, and holding it by the point with the right hand, lay it on its
face with the butt towards the left so that the bare
spot shall come at the upper end of the floss silk tip,
and pass two turns of the flying silk; insert a piece
of tinsel in the same manner parallel to and just
over the hackle, and having fastened it, hold the
tying silk with the catch; take up the dubbing of
mohair with your right hand and spin it over the
tying silk towards the left, having again taken the
latter into the right as soon as you have caught the
end of the mohair with the stop. Shape the mohair
so that the body shall taper and twist it evenly
together with the tying silk towards the shoulder,
using the stop all the way, and do not carry it too

close to the head ; pull off the superfluous mohair
with the fingers of the right hand and pass the silk
four turns over the upper end of the body, and
winding it towards the head slip it between the
gut and the hook. In this way you can always
secure the tying silk when you wish to lay down
your work. Spring the pliers on to the tinsel, and
with the right forefinger pass four even open coils
carefully and regularly; unwind the silk, and having
secured the tinsel replace it. If these coils are imperfect or irregular, neatness cannot be obtained.
Having cut off the tinsel, catch with the spring pliers
the butt of the hackle and follow the edge of the
tinsel; rolling the hackle on its back so that the
fibres shall point down the shank. When you reach
the shoulder pass several turns of the hackle close
above one another, and bringing down the tying silk
secure the butt. If one hackle is not sufficient, and
it rarely is, introduce a new hackle close above the
first, precisely as you did the other, only on its back,
and wind a sufficient number of close coils and again
fasten it. The second hackle, if weak, may be fastened in on its back by the butt, and wound with the
point.

The silk being hitched under the gut cut it off and
apply a new piece as you did the second, and wind
it towards the shoulder, letting it hang close down
to the hackle. Prepare the wings by cutting with
a sharp knife a few fibres from each of two mated
feathers, together with a little of the stem, so that
the fibres shall not be separated, and taking one

piece by the butt in the right hand, lay it on the side
of the hook next to you, and holding it with the left
pass two turns securely, but not so tight as to de-
range the feather; then catching the silk, pull the
butt fearlessly into its proper place, and passing
another turn firmly, hitch the silk under the gut, and
bring it over the reversed way on top of the wing.
Cut off the butt and taking the hook in the left hand
with the head towards the left, apply and hold the
other wing with the right hand. Still keeping the
hook reversed and wind two turns of silk with the
left hand from you, and having arranged the butt
pass another turn and hitch the silk again under the
gut, so as to reverse it for the second time. If the
wings are in their proper place, equally on each
side of the hook, restore the latter to its original
position in the left hand, and having cut off the butt
neatly, pass as many turns as you think advisable;
then having with your nails stripped off the fibres
from the butt end of an ostrich herl, tie it in with the
point towards the left and the elevated ridge of its
stem above. Hitching the thread again under the
gut, wind with the spring pliers the herl in close
coils to form the head; secure and cut it close, and
then stopping one end of silk under your forefinger
whip the other over it three turns and draw all tight.
Apply a little varnish at the head and your fly is
finished.

To strengthen the fly, it is well to use a little varnish before the head is commenced, and even before
the wings are laid, but the writer's experience goes to

prove that the wings are the last part of the fly to give out. The head will be smaller if instead of the ordinary tying silk three single strands of floss are used.

To make a handsome fly, fasten the hook, the tag, the tip, and the tail as directed, then preparing an ostrich herl as for a head, tie it in and wind several coils close to and covering the butt of the tail, holding the hook in the right hand with the silk coiled up out of the way, and using the pliers to guide the herl. Secure the end, apply with the left hand at the nearer side of the hook, the tinsel, and afterwards at the further side floss, for the body. Coil the tying silk out of the way, and with the left hand wind the floss half way up the shank and secure it; then tie in a hackle and some dubbing as heretofore directed, and having spun the latter on the tying silk with the right hand, work it up towards the head for the second division of the body, and secure it firmly. Hitch the silk under the gut, and thrusting the butt of the hackle down through the gut loop, with the pliers sprung on to the tinsel, and on the left forefinger coil the tinsel up as far as the hackle; withdraw the latter from the loop, hold it and the hook in the left hand, and with the right forefinger continue the tinsel to the head. Secure it; wind and secure the hackle as heretofore, and apply a new piece of tying silk composed of strands of floss.

Select a few fibres of various feathers, which, combined, will produce a pleasing effect, and holding

FLY-MAKING.

them all together in the left hand twist the lower half, that nearest the stem several times, and break it with the nails of the right thumb and finger, till the fibres are softened at the spot where they are to be tied to the hook. Include with them a piece of herl, and applying them with the right hand to the hook, hold them and it with the left, while you take sufficient turns of silk with the right, hitch the silk and springing the pliers on to the herl, wind and fasten the head and finish off.

There may be as many joints or divisions as fancy shall dictate; and they can be either of floss silk, mohair, or other material. To conceal the joints herl may be wound like a head or a few turns of hackle taken, or two small feathers from the golden pheasant's neck may be applied, one above and the other below, and after being loosely tied they may be drawn down by the butts till they are separate round the entire joint. The favorite feather for the tail is the golden pheasant top-knot, but in many flies scarlet worsted is preferable, and the fibres of other feathers may be substituted. In making a mixed wing as it is called, separate the fibres as much as possible, and after the wing is fastened, a long golden pheasant top-knot tied over it will often improve the effect. It is common to add to the wing two fibres of blue macaw, one on each side, and to tie them properly the silk should be reversed by passing it under the gut, as directed for tying simple wings. Care and experience are requisite to the selection of a handsome mixed wing, and fibres ɔf mallard or wood duck, plain or dyed, are usually a component part. Delicate feathers produce a finer effect than coarse ones.

In tying in an entire plume reduce it to the proper size by pulling off the fibres, and if the stem is large pare it away and always flatten and work it with the nails; then tie it loosely till it is properly arranged, and finally, secure it with a number of turns. It will slip unless made unusually firm, which the smallness of the head will readily permit.

Where the tail is worsted, it may be made of several thicknesses, left longer than necessary, and pared down and picked out after the fly is finished. As it is essential that in making a head, the ridge of the stem of the herl should be above, and as it is often obstinate in its refusal to take that position, it may be wound either way,—that is, from you or towards you.

Care should be taken with simple wings that each is in the same relative position to the body, and that the fibres are not separated; with this object not only must the thread be reversed as above directed, but cohesive feathers should be selected. Some are exceedingly difficult to tie, while others, such as the pheasant and turkey, retain their place readily. They should be selected from feathers taken from the opposite sides of the bird; and if two or more different kinds are to be used, the first wing should be completed before the other is commenced, and before the thread is reversed.

In rolling an ordinary feather in place of a hackle,

the same course may be taken as with the latter, but the better way where it is large enough is to strip off the fibres of one side, and then pare away the stem with a sharp knife. This requires care lest the knife slip and cut your hopes in twain. The same may be done with a simple hackle where great neatness is required, except that the stem does not need paring.

The tinsel may be double, tied in on opposite sides of the hook and wound contrary ways, but the effect is hardly better than a simple twist. In the latter avoid too many coils ; they should not exceed four on hooks numbered not larger than one and a half.

Two hackles, which, if the colors are well contrasted, produce a fine effect, are usually rolled together, but may be wound one after the other if care is taken to pick out the fibres. They are tied in at one time and handled as though they composed but one.

A trout-fly may be made in the manner heretofore directed for salmon-flies, omitting as much as you please, or the wings may be laid together back to back or face to face, held in that position in the left hand, and applied to the hook after the fibres have been pinched with the nails at the proper place. Being secured in that way they resemble the wings of the *ephemeræ* closely ; whereas to make one of the *phryganidæ* a few fibres of one side may be stripped off and tied on alone, lying close down upon the hook. Remember the *ephemeridæ* have whisks, the *phryganidæ* have none ; the wings of the former stand up, of the latter lie down. Coarse fibres of hackle, or golden pheasant breast and back, are usually employed for whisks ; and two strands of floss carefully waxed with a small edge of the wax, will make a tying silk as strong and large as should be used for a small fly. If well waxed, the finer the silk the firmer it holds; if not waxed no silk whatever will hold.

Another way of tying a trout-fly, by which more life is supposed to be given to it, is by commencing to fasten the gut at the bend and finishing at the head, holding the hook reversed ; then change the hook to its proper position, and reversing the thread, lay on the wings, which are composed of two strips of feather folded, so that they shall point up along the gut ; secure them firmly and cut off the butts close, divide them with the point of the picker and pass the thread through the opening each way several times, and if necessary above them both, but not on the root of the wings, till they stand up, then pushing them into their original position tie in below them by the larger end a hackle and a piece of round tinsel, and spinning a little dubbing on the silk, wind it toward the bend ; hold the thread with the catch, and with the pliers wind the tinsel and afterwards the hackle, and fasten both at the bend ; and finish off with two half-hitches. The silk composing the material in which the round tinsel is wound may be left for a tail, the coating being pulled off; or the tip of the hackle may be so left, or proper whisks may be introduced. The wings being drawn into their appropriate place will remain there, and offering resistance to tne water are supposed by some to imitate motion. Those tied in this manner are not handsome, but are great favorites with certain fishermen for their assumed killing qualities, and are considered ruined if the silk covers the roots of the wings, as is done by most Irish flytiers.

Flies may also be finished at the shoulder under the wing; a course that seems to offer no advantages and to combine most disadvantages. Or the body may be tied, beginning at the shoulder and finishing at the bend, as last described, omitting the wings and leaving a place for them till the last; a new piece of thread is then applied, and the wings being tied in their natural position, the second finish is made at the head.

To prepare two single strands of floss as tying silk, hold one end between your teeth, twist the silk and rub it lightly with a small edge of wax. If the weather is cold the wax may require thumbing before it can be used or will stick to the silk. There will be found considerable difference in the strength of strands of floss according to the color, and in very small flies this may be suited to the insect intended to be imitated, and the necessity of any other body avoided.

The word buzz, which is taken from the buzzing motion of an insect's wings when moved rapidly, is applied to the hackle wound more or less along the body, and supposed thus without wings to represent that motion. The hackle may be carried all the way from the bend or only part of the way, or merely tied very full at the head. In this matter, as well as concerning palmers, writers differ. A palmer is properly a long-bodied fly with two small hooks, and hackles wound the entire length, to represent a caterpillar and its hairy ornaments. The hooks are often made double expressly for this purpose. A hackle has but one hook and a shorter body. The word midge is another word that leads to mistakes; there are only a few proper midge-flies, such as the gnat, ant, etc., but any fly may be dressed on a minute hook and called a midge-fly, although this is not an accurate use of language. Horse-hair is sometimes used as a substitute for gut by old-fashioned anglers, but it is weaker, more apt to slip, and more perceptible to the fish.

An excellent plan for preserving feathers conveniently and safely, is to put them in envelopes suited in size to their length, and to stow them, together with a piece of camphor, in a tin box. If they are looked over occasionally, and the camphor renewed as it wastes, they will remain untouched by moth; but if they are to be kept for a long time unhandled, they should be deposited in a linen bag. The envelopes should be large, for if the fibres are bent they will not make handsome wings, and the different classes of feathers may be tied in separate bundles.

The following wax is recommended in the Appendix to "Fly-fishing in Salt and Fresh Water:"— Melt some resin in a small vessel over a slow fire, and whilst it is on the fire and after it has become

fluid, take a pure white wax candle, light it and let
it drop into the melted resin ; there is no rule as to
the quantity. Pour out upon a board either greased
or rubbed with wax from the candle, one fourth of
the composition ; then drop more wax into the re-
mainder and pour out one fourth more. Proceed in
the same manner with the other two fourths, and
thus you will have wax of four degrees of hardness ;
that with the least wax dropped from the candle
being for use in hot weather, the others for different
degrees of temperature of the seasons. After the
composition has become cool on the board, it should
be well worked on the board as shoemaker's
wax is.

To make soft wax to use upon very delicate silk,
dissolve some common shoemaker's wax in spirits of
wine until it becomes of the consistency of butter,
then put a small quantity on the inside of a piece of
an old kid glove, and draw the silk gently through
it. Or put a piece of shoemaker's wax the size of
a walnut in a small bottle, and pour over it an ounce
of eau-de-cologne ; shake it occasionally till it dis-
solves, when it is ready for use ; then taking a drop
between the finger and thumb, draw the silk through
it. It may be carried in a metal bottle with a
screw stopper, and if well corked will keep for
years.

In Scrope's Days and Nights of Salmon Fishing,
is found the following description of a few favorite
salmon flies :—

No. 1. Kinmont Willie.

Wings.—Mottled feather from under the wing of
a male teal.
Head.—Yellow wool.
Body.—Fur of the hare's ear.
End of Body.—Red wool.
Tail.—Yellow wool.
Round the body.—Black cock's hackle.

No. 2. Lady of Mertoun.

Wings.—Mottled feather from under the wing of
the male teal.
Head.—Crimson wool.
Body.—Water rat's fur.
End of body.—Crimson wool.
Tail.—Yellow wool.
Round the body.—Black cock's hackle.
End of body.—A little red hackle.

No. 3. Toppy.

Wings.—Black feather from a turkey's tail tipped
with white.
Head.—Crimson wool.
Body.—Black bullock's hair.
End of body.—Crimson wool.
Tail.—Yellow wool.
Body.—Black cock's hackle.
End of body.—Small piece of red cock's hackle.

No. 4. Michael Scott.

Wings.—Mottled feather from the back of a drake (mallard).

Head.—Yellow wool with a little hare's fur next to it.

Body.—Black wool.

End of the body.—Fur from the hare's ear; next to the hare's ear, crimson wool.

Tail.—Yellow wool.

Round the body.—Black cock's hackle.

End of the body.—Red cock's hackle.

Round the body.—Gold twist spirally.

No. 5. Meg with the Muckle Mouth.

Wings.—From the tail of a brown turkey.

Head.—Crimson wool.

Body.—Yellow silk.

End of body.—Crimson wool.

Tail.—Yellow or orange wool.

Round the body.—Red cock's hackle.

Round the body.—Gold twist; over it hackle mixed with color, as above.

No. 6. Meg in her Braws.

Wings.—Light brown from the wing of a bittern.

Head.—Yellow wool.

Next the head.—Mottled blue feather from a jay's wing.

Body.—Brown wool mixed with bullock's hair.

Towards the end of body.—Green wool; next to that crimson wool.

Tail.—Yellow wool.

Round the body.—Gold twist; over that cock's hackle, black at the roots and red at the points.

"Concerning these flies, I will note one thing, which is, that if you rise a fish with the Lady of Mertoun, and he does not touch her, give him a rest and come over him with the Toppy, and you have him to a certainty, and *vice-versâ.* This I hold to be an invaluable secret, and is the only change that, during my long practice, I have found eminently successful.

Another method of dressing No. 3, Toppy; wing feather from rump or tail of turkey, which is black below and strongly marked with a white tip, to be set on Tweed fashion (that is to say, the wings parted and made to lie open like a butterfly's wings).

"Body black mohair; three turns of broad silver tinsel.

"Blue or black heron's neck-feather at the shoulder; if heron's feather cannot be procured, a good-sized black cock's hackle; orange or yellow wool, for tail."

The long transparent bodies which are made in imitation of the *ephemeridæ,* and are rather more admired by the fancy angler than by the fish, are composed of small pieces of gut, whalebone, or other similar material, which, after being cut to the proper length, are fastened on at the shoulder, together with a thin flat end of gut, such as comes in the covered part of every hank, and which, after being well soaked in warm water, has been smoothed down with the finger nail. The latter, while still damp and pliable, is wound evenly round the material of

10

the body, including the hook, for several turns, and
then round the body alone, and secured at the extremity by passing a couple of turns over the end
and drawing it through. As this is transparent, it
will show the color of the substance below, and may
even be wound over floss-silk bodies which do not
project beyond the hook, and while adding brilliancy,
will protect them from injury. The whisks may be
included with the solid material of the body, and
an upper section may be added ; the hackles are to
be introduced, and the wings secured afterwards ;
but although a very perfect imitation, it is not generally so killing as the ordinary artificial fly.

In giving the preceding directions, it is by no means
intended to advise that the table vice should be
discarded ; but, on the contrary, a small or handsome fly can be tied much more easily with its assistance. A little practice with the fingers alone will,
however, greatly increase one's expertness, and remove an awkward difficulty in case the vice should
by any chance be left behind. The great objection
to tying a fly with the fingers is the risk of mussing
the feathers, especially in summer, when perspiration
prevails.

I am indebted to Mr. J. James Hyde, a gentleman who, although an amateur, is one of the most
finished anglers and neatest dressers of a well-imi-
tated trout-fly in the United States, for the following directions for tying all Ronalds's flies with the
feathers of our American birds, so that the angler
who may be unacquainted with the English feathers

can make an accurate imitation, and not, as is too common in this country, produce some wretched abortion for a well-known fly, and may at the same time avoid the unnecessary outlay of importing expensive foreign materials.

The following list of flies is taken from Alfred Ronalds's "Fly-Fisher's Entomology." This work has been selected because its descriptions are imitations of real flies, and not of traditional or conventional nondescripts, which, although the delight of professional dressers, might be safely worshipped without breaking the commandment, since they are not the "likeness of any thing in the heaven above, nor in the earth beneath, nor in the waters under the earth."

Some alterations have been made for the purpose of facilitating the reader in his choice of materials, and the feathers indicated are, in most cases, those of our own birds, which may be easily procured, and are quite as suitable as the foreign ones given by Ronalds. Mohair is the best material for the bodies of trout-flies, and though others are sometimes named as being an easier method, the experienced amateur will prefer mohair, with which he will produce the same effect, without any of the objections to which all other materials are liable; and by a judicious mixture, any shade of color may be obtained.

Ronalds's work being descriptive of English flies only, it has been deemed advisable to substitute their American prototypes in all cases where they are known; and although the trout are not perhaps thorough entomologists, the scientific fisherman will always prefer to use a fly which exists in the waters he frequents, to an English resemblance, restricted perhaps to a confined locality some thousands of miles away. As a general rule, there is no doubt that the best imitations of the fly the fish are taking will be the most successful; yet there are exceptions, of which the ibis fly is a *glaring* instance. It is also desirable at times to vary the sizes of flies, and to make the imitations larger than the living flies—when, for instance, the water is rough or thick; but these variations are not of absolute importance.

No. 1. The Blue Dun.

This fly is the earliest American *ephemera*, and may be found on warm days in February. In March it is abundant. It lives three or four days, and then becomes the red spinner.

Imitation.

Body.—Mouse-colored mohair, spun very thinly on yellow silk.

Tail.—Two fibres of gray mallard.

Wings.—From a quill-feather of the robin's wing. The third or fourth feather with a tinge of reddish brown at the extremity of the fibre.

Legs.—Two or three turns of a blue or ginger dun hackle. One side of the hackle may be stripped off for the *ephemeridæ*.

No. 2. The Red Spinner.

This is the blue dun in its perfect or *imago* state.

It is now of a reddish brown, and its wings are near-ly transparent. It lives four or five days, but if the weather be hot, will be found more at even-ing.

Imitation.

Body.—Of bright reddish brown mohair, ribbed with silk of same color.

Tail.—Two whisks of a red cock's hackle, or of the red body-feather of the golden pheasant.

Wings.—From a thin, transparent mottled grey feather of the mallard or wood-duck.

Legs.—Plain red cock's hackle. The wings of the *ephemeridæ* stand upright on their backs.

No. 3. The Water Cricket.

This insect lives upon small flies, etc., whose blood it sucks in a manner similar to that of the land spider. It runs upon the water and darts upon its prey while struggling on the surface. In the sum-mer months it is provided with wings.

Body.—Orange mohair, spun on black silk, and ribbed with black silk.

Legs and Wings.—A black cock's hackle. This fly is always made buzz. The wings are very trans-parent.

No. 4. Great Dark Drone.

This fly is found upon the grass in a torpid state, until the sun warms the air, when it takes wing; and afterwards, if there be a breeze, it is found upon the water. They are of great variety of color, but the black is the most common.

Imitation.

Body.—Black mohair spun thickly on black silk

Wings.—The dun feather of a mallard wing. The wings lie flat upon its back, and the upper fibres of the hackle should be cut off.

Legs.—A dark grizzled hackle. This is a late fly.

No. 5. Cow-Dung Fly.

This fly is to be found throughout the year. It is most abundant in March, and during a high wind it is blown upon the water. The color of the male is a tawny yellow; that of the female a greenish brown.

Male.—Imitation.

Body.—Yellow and light-brown mohair mixed, spun on light brown silk.

Wings.—The wing feather of the brown thrush, or of the rail (corncrake).

Legs.—A ginger-colored hackle.

Female.—Olive-colored mohair body; wings and legs the same. The wings lie flat, and the upper hackles should be cut off.

No. 6. Peacock Fly.

This is a small beetle, very abundant on warm summer days. It often falls upon the water in its flight, or is blown upon it by the wind. It is highly praised by English writers, and is described by *Arundo*, in "Practical Fly-Fishing," as "the little chap."

Imitation.

Body.—Copper-colored peacock's herl.

Wings.—The darkest part of a robin's wing-feather.

Legs.—A dark purple-dyed hackle.

No. 7. MARCH BROWN.

This *ephemera* is the next in season after the blue dun. It is a handsome and attractive fly, and is eagerly devoured by the trout. The male is of a chocolate color, and the female a greenish brown. It lives three or four days, and then changes into the great red spinner.

Imitation.

Body.—Sandy-brown mohair, ribbed over with olive silk.

Tail.—Two fibres of a brown hen's feather.

Wings.—From the mottled wing-feather of a brown hen, which may be found of the exact shade.

Legs.—A brown hen's hackle, or the small brown body-feather of the widgeon.

No. 8. GREAT RED SPINNER.

This is the *metamorphosis* of the March brown, and may be used on warm evenings through the season. It is a very excellent fly.

Imitation.

Body.—Orange and brown mohair mixed, ribbed with fine gold twist.

Tail.—Two fibres of a bright amber red hackle, or the body-feather of the golden pheasant, which is a strong, durable feather for this purpose, and may be found from a bright yellow to deep red.

Wings.—Light-colored feather from the robin's wing.

Legs.—A bright amber red hackle.

No. 9. Sand Fly.

This fly comes from a water *larva*, and is one of the best flies which can be used during April and May. Its wings are long and full, and lie flat upon its back.

Imitation.

Body.—Sandy-colored mohair, spun on silk of the same color.

Wings.—From the wing-feather of the brown thrush, or the mottled brown feather of a young hen.

Legs.—A light ginger hackle. Cut off the upper fibres of the hackle, that the wings may lie flat.

No. 10. The Stone Fly.

This fly also comes from a water *larva*. It is heavy in its flight, but runs with great rapidity, and is generally found in streams, amongst the stones or close to the sides of the water. Its body is nearly half an inch in length.

Imitation.

Body.—Brown and yellow mohair mixed, and ribbed with yellow silk.

Tail.—Two strands of brown hen's wing.

Wings.—From the mottled feather of a brown hen made full, and to lie flat.

Legs.—A grizzled hackle.

No. 11. The Gravel Bed, or Spider Fly.

This fly is found only in running waters, but where it is found it is very numerous. It may be used all day, and is a very delicate fly. It will raise fish in clear water when no other fly will.

Imitation.

Body.—Lead-colored silk thread, with which the fly is tied. Fine and thin.

Wings.—From an under covert feather of the wood-cock's wing. To lie flat.

Legs.—Two turns only of a black hackle.

No. 12. The Grannom, or Green Tail.

This fly comes from a water *larva*, and is found chiefly at morning and at evening. The green tint of its body is derived from the color of the bag of eggs near the tail. There are a number of species in the United States, and in some the bag of eggs is yellow, and in some orange. The green is the most used.

10*

Imitation.

Body.—Work in a little tuft of green at the tail, and then finish the body of sandy-colored mohair.

Wings.—A light brown mottled hen's feather, to lie flat.

Legs.—A pale ginger hackle,

The body of the male is yellow, without the green tag.

No. 13. The Yellow Dun.

This beautiful *ephemera* is one of our very best flies. There are several varieties, and some of them are an inch in length. It changes to a spinner, very similar to the metamorphosis of the blue dun (No. 2), only lighter and yellower, and should be so tied.

Imitation.

Body.—Yellow mohair spun very thinly on pale blue silk.

Wings.—From the lightest part of the feather of a robin's wing.

Legs.—A pale yellow dun hackle.

This fly must not be finished off at the head with the blue silk, but a yellow must be tied in for the purpose when the body is done.

No. 14. The Iron Blue Dun.

This is one of the smallest of the *ephemeridæ*, but not the least useful. It lives only two or three days before changing its coat, when its body becomes almost white, and its wings transparent.

Imitation.

Body.—Pale blue mohair, very thinly spun on reddish-brown silk, with which the head must be finished.

Tail.—Two whisks of the yellow body-feather of the golden pheasant.

Wings.—From the wing-feather of the blue-bird.

Legs.—A very small yellow dun hackle.

No. 15. The Jenny Spinner.

This is the name of the iron blue dun (No. 14) in his new dress, in which he lives four or five days. It is a killing fly towards evening in clear water in summer. There are in the United States at least some hundred varieties of these small *ephemeridæ*, of every conceivable color, and the skilful dresser will take pleasure in tying them, using the feathers of the small domestic and foreign birds which he can procure. Such are the sky-blue, the orange dun, the pale evening dun, the July dun (blue and yellow), the whirling blue dun, and the little pale dun.

Imitation.

Body.—White floss silk, tied at head and tail with brown silk thread.

Tail.—Two whisks light dun hackle.

Wings.—From a blue-bird's wing-feather

Legs.—A very small and very light dun hackle, nearly white.

No. 16. The Little Yellow May Dun.

This is another of the *ephemeridæ*, and a most useful one to the fisherman. It is not so small as the preceding one (No. 14), and changes to a very light red spinner.

Imitation.

Body.—Pale ginger-colored mohair, ribbed with yellow silk.

Tail.—Two whisks of yellow, or ginger hackle.

Wings.—Mottled feather of the mallard, dyed a greenish yellow.

Legs.—Light ginger hackle, dyed the same color as the wings.

No. 17. The Black Gnat.

Every fisherman is familiar with this little insect, and has taken trout with their mouths and throats filled with them. It is, however, not properly a gnat, but a midge.

Imitation.

Body.—Black ostrich herl.

Wings.—The darkest feather of a robin's wing.

Legs.—A black hackle.

The black midge should be made similarly, but with a *thin* black silk body.

No. 18. The Oak Fly, also the Down Head Fly, and Down Hill Fly.

This is a land fly, and may be found upon the

trunks of trees or on posts near the water. It is carried on the water by the wind, and is consequently used with most success on windy days, like the cowdung.

Imitation.

Body.—Orange floss silk or mohair, ribbed with black silk.

Wings.—The darkest part of the wing-feather of a curlew.

Legs.—A furnace, or red and black hackle.

No. 19. The Turkey Brown.

This *ephemera* is common to most of the waters of New York, and is found on nearly all the Long Island ponds, where it is eagerly taken by the trout. It appears about the middle of April, and changes to a little dark spinner, which is a most killing fly just before dusk.

Imitation.

Body.—Brown mohair ribbed with purple silk. The female is of a *greenish* brown.

Tail.—Two fibres of the same feather as the wings.

Wings.—Of the brown mottled feather from the back of a ruffed grouse.

Legs.—A red-brown hackle.

No. 20. The Little Dark Spinner.

This is the perfect, or *Imago*, state of the turkey brown (No. 19) just described. It is as fragile as it is beautiful, and can hardly be touched without maiming or killing it.

Imitation.

Body.—Light reddish-brown floss silk, ribbed with purple.

Tail.—Three whisks of a light dun hackle.

Wings.—From a feather of the robin's wing, or the under feather of a young grouse's wing.

Legs.—A light dun hackle.

No. 21. THE YELLOW SALLY.

This is a water fly, which continues in season for four or five weeks from the middle of May. Its wings are transparent, and lie close and flat. It is sometimes called "the flat yellow."

Imitation.

Body.—Yellow mohair, ribbed with pale green silk thread.

Wings.—White pigeon wing, stained a pale greenish yellow.

Legs.—A white hackle, dyed the same color as the wings.

No. 22. THE FERN FLY.

The two most common varieties of this fly are known as the "Soldier" and the "Sailor." The wing coverings of one are red, and of the other blue. They are both well taken by the trout until the end of July, on hot days.

Imitation.

Body.—Orange floss silk.

Wings.—The darkest part of a robin's wing-feather.

Legs.—A red cock's hackle.

Two or three fibres of some blue feather may be tied in with each wing, on the outside, or of red, to represent the wing-covers.

No. 23. The Alder Fly.

This fly comes from a water *nympha.* It lays its eggs upon the leaves of trees which overhang the water, whence they drop into it. It is in season during May and June.

Imitation.

Body.—Peacock's herl tied with black silk.

Wings.—From a feather of a brown hen, made large and full.

Legs.—A black cock's hackle.

No. 24. The Green Drake.

This is the most famous of all the English *ephemeridæ.* It is a large and beautiful fly, but is not found, so far as known, except in running waters. For ordinary streams and ponds here the "little yellow May dun" (No. 16) will be found preferable.

Imitation.

Body.—Straw-colored floss silk, ribbed with brown ; the head of peacock's herl.

Tail.—Three hairs from a fitch's tail.

Wings.—From a mottled feather of the mallard, stained a greenish yellow.

The female of this fly changes to the grey drake, and the male to the black drake. They are little used.

No. 25. The Hazel Fly.

This is a beetle, the *pupa* of which inhabits the earth. It is found upon poplar-trees, and a species very similar is found upon fern. It is blown upon the water, and is to be used on windy days.

Imitation.

Body.—A black ostrich herl and a peacock's herl, twisted together on red silk.

Wings and Legs.—Made buzz with a dark furnace hackle.

As this fly never alights upon the water, it is generally seen struggling with its wings in motion.

No. 26. The Dark Mackerel.

This is the *imago*, or perfect state of another kind of green drake, darker than No. 24. It is found in some waters where the true green drake is not, and is used in its stead.

Imitation.

Body.—Dark mulberry floss silk, ribbed with fine gold twist.

Tail.—Three hairs from a fitch's tail.

Wings.—From the brown mottled feather of the mallard, which hangs from the back over a part of the wing.

Legs.—A dark purple hackle.

No. 27. The Gold-Eyed Gauze Wing.

This beautiful insect is not found upon all waters, but where it is, affords great sport on windy days. It may be used from June till the end of September.

Imitation.

Body.—Pale yellowish green floss silk, tied with silk of the same color.

Legs.—Pale blue dun hackle, with one or two turns *in front of the wings.*

Wings.—A pale transparent mallard, or wood-duck feather, stained slightly green. Very full, long, and to lie flat.

No. 28. The Wren Tail.

This is a species of *hopper*, sometimes called " *ant hoppers.*" They hop and fly for about twenty yards, and sometimes drop short and fall upon the water. The light and dark brown, and the greenish blue, are the most common.

Imitation.

Body.—Ginger-colored mohair ribbed with fine gold twist, short.

Wings and Legs.—Feather from a wren's tail, wound on hackle-wise.

A brown mottled hackle may be used in place of the wren's tail feather.

No. 29. THE RED ANT.

There are many species of these winged ants, and they are familiar to every one. The red and black are those generally used.

Imitation.

Body.—Copper-colored peacock's herl, wound thickly, for two or three turns, at the tail to form a tuft; the rest of the body dark red silk.

Wings.—From the lightest part of a robin's wing. To lie flat.

Legs.—A small red hackle.

The black ant is made of black ostrich herl body; wings from the darkest part of a robin's wing; legs, a small black hackle.

No. 30. THE SILVER HORNS.

This fly is an excellent one until the end of August, principally in showery weather.

Imitation.

Body.—Black ostrich herl tied with black silk, and trimmed down.

Wings.—A wing-feather of the black-bird.

Legs.—Small black cock's hackle.

Horns.—Two strands of the grey feather of the mallard.

The male has black horns. To make it buzz,

the body is to be ribbed with silver twist upon the black ostrich herl, and a black hackle wrapped the whole length of the body.

No. 31. The August Dun.

This fly comes from a water *nympha*, lives two or three days, and changes to a red spinner. This fly is for August what the March brown is for March.

Imitation.

Body.—Brown floss silk, ribbed with yellow silk thread.

Tail.—Two hairs from a fitch's tail.

Wings.—Feather of a brown hen's wing.

Legs.—Plain brown hackle.

Made buzz with a grouse feather, in place of wings and legs

No. 32. The Orange Fly.

This is an Ichneumon Fly. It is furnished with an *ovipositor*, for the purpose of piercing the skins of caterpillars, in which it deposits its eggs, the grub from which grows in, and ultimately kills, the insect in which it was hatched.

Imitation.

Body.—Orange floss silk tied on with black. Thick and square at the tail.

Wings.—Darkest part of a robin's wing.

Legs.—A very dark furnace hackle.

No. 33. The Cinnamon Fly.

This fly comes from a water *pupa*. It should be used after a shower, and on a windy day. It is a very killing fly on some waters, and somewhat resembles the land fly, but does not appear so early.

Imitation.

Body.—Fawn-colored mohair, tied on silk of the same color.

Wings.—Feather of a yellow-brown hen's wing, rather darker than the thrush feather. To lie flat.

Legs.—A ginger hackle.

The pinnated grouse's small wing-feather, dyed a pale cinnamon with madder and copperas, is an excellent feather for the wings of this fly, and of No. 34.

No. 34. The Cinnamon Dun.

This *ephemera* is found in abundance on the streams in Pike Co., Pa., and in some other localities. It is similar to the little yellow May dun, but is of a bright cinnamon color, and comes on in July and August. Its *metamorphosis* is of a light red brown, with wings almost white.

Imitation.

Body.—Red and yellow mohair spun on yellow silk, and ribbed with the same.

Wings.—The light feather of a grouse's wing, dyed cinnamon with madder, or the feather of a curlew's wing.

Tail.—Two fibres of the same feather as the wings.

Legs.—A ginger hackle.

No. 35. The Blue Bottle.

This and the house fly become blind and weak in September, are frequently blown upon the water, and afford good sport. They may be used especially after a frosty night, but are not unsuccessful earlier in the season.

Imitation.

Body.—Bright blue mohair, tied with light brown silk. The body thick.

Wings.—The lightest feather of a robin's wing.

Legs.—Two turns of a black hackle.

The *House Fly* may be made thus:

Body.—Light brown and green mohair mixed.

Wings.—Light-colored feather from a robin's wing.

Legs.—A blue dun hackle.

Head.—Green peacock's herl, with two or three turns under the wings.

No. 36. The Red Palmer.

This is the caterpillar of the garden tiger-moth. This palmer is found early in the spring, and is chiefly recommended for streams where trees overhang the water. Cuvier states that this caterpillar changes its skin ten times during its growth.

Imitation.

Body.—Peacock's herl, with a red cock's hackle wrapped the whole length, and tied with red silk.

Ronalds's palmers are made long, and have a second hook tied in about half way up the body. It is a killing fly in streams, and of little use in ponds in the United States.

No. 37. The Brown Palmer.

The preceding remarks on the red palmer apply equally to this and the succeeding description. The white and yellow are equally successful on wooded streams, and they all may be used through the season.

Imitation.

Body.—Light brown mohair spun on brown silk, and a brown cock's hackle wrapped all the way up.

No. 38. The Black and Red Palmer.

Imitation.

Body.—Black ostrich herl, ribbed with gold twist, and a red cock's hackle wrapped over it.

The feather at the shoulder should be a large furnace hackle, and the herl should be thickest there. Show the gold twist clearly at the tail.

THE ART OF DYEING FEATHERS, HACKLES, PIG'S WOOL, AND MOHAIR, SUITABLE COLORS FOR FLIES.

It is a great advantage to the fly-fisherman to possess the knowledge of dyeing his materials, as it is by no means easy to procure them at all times of the desired color. It is, besides, an amusement and an inducement to study the colors, sizes, and habits of the insects which he wishes to imitate. The colors for salmon-flies should be as rich and brilliant as possible; those for trout are of soberer hues. Hackles should be selected with much care, of fine fibre, of even taper. White hackles are requisite for yellow, orange, blue, and green; red hackles for claret, red, brown, and olive. They should be washed in soap and water before dyeing, and tied in small bunches for convenience of handling.

It is important in dyeing all kinds of feathers to dress them thoroughly. They should be rinsed in clean water when taken from the dye, wiped as dry as possible, and dressed with the hand in the direction of the fibres until dry. This gives them a smoothness and gloss which can be given in no other way.

Naturally-colored feathers are perhaps preferable, as a general thing, for trout-flies; but there are some which cannot be had of the proper color, and for salmon-flies the dyer's art is indispensable.

To Dye Yellow.

Put two table-spoonfuls of ground alum, and one tea-spoonful of cream of tartar into a pint of water. When perfectly dissolved and boiling, put in the feathers, hackles, or hair, and simmer for half an hour. Take them from this mordant bath, and put them in the yellow dye, made by infusing a table-spoonful of ground turmeric in a pint of water, and immersed until the color is extracted.

Boil until the color is deep enough, and then wash them in clean water. Dry, and dress them as directed.

There are several materials for yellow dyes, such as fustic, quercitron bark, yellow wood, Persian berries, and weld; but turmeric is the best for the purpose.

To Dye Orange.

To produce orange the feathers or other material should be first dyed yellow, according to the previous recipe. They should then be boiled in a dye made with madder and a small quantity of cochineal, until the requisite shade is obtained.

To Dye Scarlet.

Make a strong infusion of cochineal, put in a few drops of muriate of tin, which will make a crimson, and then put in a little cream of tartar, which will make a clear scarlet. The proportions in weight are one part of muriate of tin to two parts of cream of tartar. It is best to boil the feathers first in the solution of alum. Simmer them until the color is obtained.

To Dye Crimson.

Boil the materials to be dyed in a solution of alum and cream of tartar, for half an hour; bruise two table-spoonfuls of cochineal, and simmer them in water until the color is extracted.

Take the materials from the alum water, and boil them in the cochineal liquor until you have the color you wish.

Wash them in clean water, and if feathers, dress them until dry.

To Dye Brown.

Brown may be procured by boiling walnut shells to a strong solution, and of a more chestnut hue by boiling in a bath composed of a small handful each of sumach and alder bark, boiled in half a pint of water, with half a drachm of copperas.

To Dye Blue.

Boil your material in the solution of alum and tartar already described.

Then make a blue dye by dissolving the prepared indigo paste in water, the quantity of which must depend upon the color you wish to produce. Boil until you have the shade you desire.

The prepared indigo paste is made by dissolving indigo in oil of vitriol and water in a well stoppered

11

bottle, but it is some trouble to prepare, and may be had already made at a dyer's.

It requires a white ground to produce a good blue.

To Dye Purple or Violet.

First dye your materials blue and let them dry, according to the recipe already given. Then bruise a couple of table-spoonfuls of cochineal, which boil until the color is extracted; then put in the blue hackles, or other feathers, and simmer them over the fire until the purple is obtained.

Wash and dress as before directed.

To Dye Claret.

Bruise a handful of nutgalls and boil them half an hour, with a table-spoonful of oil of vitriol in half a cup of water. Put in your material and boil for two hours; add a piece of copperas the size of a walnut, and a little pearl ashes. Boil until a fine bright claret is produced.

Wash and dress as before.

To Dye Black.

Boil two handfuls of logwood with a little sumach and elder bark for an hour; put in the hackles or feathers, and boil very gently. Put in a little bruised copperas, a little argil, and some soda; leave the feathers in for some hours with a gentle heat, then wash the dye well out of them, dry and

dress them. The argil and soda must be used sparingly.

To Dye Lavender, or Blue Dun.

Boil ground logwood with bruised nutgalls and
a little copperas. The shade of color may be varied
by using more or less of the materials.

You may have grey, and duns of various shades,
by boiling with the logwood a little alum and copperas.

To Dye Green.

Dye your material a light shade of blue first,
according to the directions for that color; then put
them into the yellow dye, and examine them frequently while boiling to see that you get the proper
shade. You may get any shade of green by dyeing
the blues darker or lighter, and then boiling them a
shorter or longer time in the yellow dye.

The blue and yellow dyes may also be mixed to
produce any shade of green, but this requires judgment and considerable experience, and the result is
not superior. It must be remembered that the blue
becomes developed by time, and the color should be
at first more yellow than is required.

To Dye a Mallard's Feather for the Green Drake, and Little Yellow May Dun.

Boil the feathers in the mordant bath of alum
already described.

Then boil them in an infusion of fustic to produce

a yellow, and subdue the brightness of this yellow by adding copperas to the infusion.

It is better to add a *little* of the indigo paste to this dye. It gives a brighter, clearer tone of color.

To Dye Gut.

An Azure, or Neutral Tint.

1 drachm logwood,
6 grains copperas.
Immerse the gut 2½ to 3 minutes.

An Azure Tint, more Pink.

1 drachm logwood,
1 scruple alum.
Immerse the gut 3 minutes.

A dingy Olive.

1 drachm logwood,
1 scruple alum,
3 scruples quercitron bark.
Immerse from 2 to 3 minutes.

A light Brown.

1 drachm madder,
1 scruple alum.
Immerse from 5 to 6 minutes.

A light Yellow, or Amber.

1½ scruples quercitron bark,
1 scruple alum,

6 grains madder,
4 drops muriate of tin,
1 scruple cream of tartar.
Immerse 2½ minutes.

An Olive Dun.

Make a strong infusion of the outside brown leaves or coating of onions, by allowing the ingredients to stand warm by the fire for ten or twelve hours.

When *quite cold* put the gut into it, and let it remain until the hue becomes as dark as may be required.

All the above dyes for gut are to be used *cold*.

THE POTOMAC.

ARTIFICIAL BAIT AND FLY-FISHING.

In fly-fishing, a rod, and a good rod, is one of the prime requisites, upon the excellence of which depends, in a great measure, the successful exercise of the angler's skill.   An excellent rod may be made of different materials and in different manners, a choice among which will depend upon fancied, more than real superiority; but each writer has his favorites, and, if able, is entitled to give the reasons for his preference.

Fly-fishing is mainly confined to salmon and trout-fishing; for these, essentially different implements are required; for the long casts and heavy play of the former, amid the rapids and cascades of the foaming river, a stout, stiff, two-handed rod is requisite; while for the feebler efforts and shorter casts of the latter, amid the ripples of the murmuring brook, or upon the placid surface of the quiet pond, a light, single-handed rod is preferable.

The salmon-rod should be as long and strong as the muscles of the angler will enable him to wield trenchantly all day through, and should have that quick, powerful pliancy that will send the fly with or across the wind a prodigious distance.   It is ordinarily made of ash or hickory for the joints, and bamboo, on account of its lightness, for the tip.

Greenheart has lately become the favorite wood, being now almost universally employed in England, and offers, certainly, some desirable advantages; but I have not had sufficient experience with it to speak decisively of its merits. A salmon-rod should be twenty feet long; after giving the matter due deliberation, and trying to reduce every ounce of weight, I have resolved that I cannot take off an inch from twenty feet. To meet the objection that a weak, small man must, under these circumstances, either give up the fishing or the rod, I would suggest that he inure himself to the labor by practising, for his first few days upon the river, with a sixteen-foot rod till his muscles are strengthened, and then substituting one of full length and weight.

A sixteen-foot rod may be handled beautifully, will cast the fly lightly, will kill a fish delicately, but it will not enable the possessor to force his line against or across a gust of wind eddying down the bank of the stream, nor to command all the casts of a broad river with facility, neither can he strike with certainty, nor kill his fish with rapidity. Salmon rivers are usually wide, sometimes wild, broken, and impassable even for that wonderful compound of life and lightness, the birch canoe, and cannot be reached in every part except with a long line under perfect control; frequently, the very spot where the fish habit, the swirl of the current or the pitch of the cascade is beyond the limits of him of the fifteen-foot rod; and if by the utmost effort the line is cast far enough, the first eddy will slack it up and deprive the weak, pliant rod of all control over it.

Again, where the favorite pool lies close by the overhanging rock, upon some accommodating ledge of which the angler crawls prone to the earth, hiding from the sharp eye of the watchful fish, he can with a long rod jerk out the line, and twitching it over the surface, beguile the prey; while with a shorter one he might be deprived of concealment, and stand confessed a laughing-stock to the fish, dangling a useless line close to the rocky bank. If the water, the wind, or the fish are strong, the rod should be the same; although advocating gentle treatment, there are times when, I assure the reader, that vigor must be exerted, and then twenty feet are better than fifteen.

No practical working rod can be made by the removal of one or more joints and the substitution of others, to increase or diminish in length. There must be a uniform taper consonant with the length, which, in case of alteration, will be destroyed, and the rod rendered harsh or feeble. The strain will not come equally upon all its parts; it will bend irregularly, and under a sudden strain is almost sure to give way. I had a rod in which a single joint could be substituted for the butt and next joint, which broke on an average of once a day so long as it was used in that way, and until the two joints were restored.

The elasticity of a good salmon-rod is like that of steel, and by the aid of such an implement alone can the fly be propelled to a proper distance. The force must be transmitted to the tip end of the leader, and the angler must feel in casting that his rod is up to its share of the work. It must neither drag, for in that case the line follows the impulse feebly; nor be too stiff, for then no life can be imparted to the line. If the rod is weak, it cannot cast with power; if it is harsh, it cannot cast at all. It must bend, but must leap back to its place, driving the fly far ahead of it by the strong and steady impulse.

A deficiency in vigor is felt at once by the angler, as a want of proper resistance to his exertion, and will be particularly noticeable of a bad day, or in an unfavorable locality, when the rod will seem to double back and fail utterly in a weak disgusting way; while too great stiffness will go to convince the angler that he is using a bean-pole.

The single-handed trout-rod is a very different affair, much more difficult both to make and handle; coarser tools and tackle will answer for the coarser fish, but nothing less than the best material and workmanship will enable the trout-fisher to perform creditably and successfully. It must be light for fine fishing, not over ten ounces in weight; it must be the perfection of elasticity; it must have a certain strength; it must balance perfectly in the hand; in other words, it must be perfection, to attain which, requires the utmost care and the greatest skill. It is a strange fact that decidedly better fly-rods, and perhaps better salmon-rods, can be obtained in America than in England, in spite of the greater foreign experience; a result that is due mainly to our persistent effort after delicacy, and perhaps partly to the habits and size of our fish; but an English fly-rod is now regarded as a clumsy monstrosity.

Trout-rods are usually made of ash with a bamboo or Calcutta cane-tip; the latter is infinitely preferable to lance-wood, on account of its greater strength and lightness. The bamboo is split into narrow pieces the length of one joint of the cane, and being glued together, is trimmed to the proper shape. Three pieces should be used, each planed, by an instrument made for the purpose, into an obtuse angle, and fitting neatly together; if two pieces only are united, the tip will bend to different degrees in different directions.

Bamboo may also be used for the second joint, and makes a light and vigorous rod, with ash for the butt; horn-beam or iron-wood, and greenheart, have also been introduced for trout-rods, but have not come into general acceptance; lance-wood is strong but too heavy, while my decided favorite is red cedar. Rods, after they have been exposed to wet, and have endured the strain of a strong fish, or even the effort of repeated casting, will warp; they will, if they are extremely light, prove deficient in power; they are apt to be either heavy or feeble; they will, when the current or wind is strong, give to it and lose their quickness in striking; in fact, they have many defects common to one or the other of the above woods, unless they are made of cedar;

in this case they have but one fault, they are brittle. A cedar rod never warps; it springs to the hand as quick as thought to the brain; it is never slow or heavy; it cannot be kept down by the wind or the current; it is never aught but quick, lively, and vigorous; it will cast three feet farther than any other rod of the same weight, and strike a fish with twice the certainty. The wood is extremely light, but the grain is short; it never loses its life, but will snap under a sudden strain.

I once struck a salmon with an eight-ounce cedar trout-rod; it was at the basin below the Falls of the Nipisiquit, where the current of the river, rushing against the calm water of the deep pool, creates a gentle ripple. The hour was near midday, and I was catching sea-trout in that profusion with which they abound in the northern waters, when out of the ripple, a few yards beyond my reach, rose a mighty monarch of the flood, and turning over as he sank, caused a heavy surge in the tide.

My Canadian guide, an enthusiastic Frenchman, was with me, and our nerves tingled and our cheeks flushed at the sight; approaching the canoe, a long cast brought him out again, but only to miss the tiny trout-fly. Convinced that he would rise, I hastily substituted a small salmon-fly for the stretcher, leaving on the leader the two small droppers I had been using, and again carefully cast over him. Out he came, the water breaking round him and rolling away in miniature circling waves, and the foam flying from the powerful blow of his tail as he turned

down. I struck, but it was as though I had struck
a rock; he darted to the bottom, making the rod fly
in splinters; at every surge fresh splinters broke off
and fell about in showers; a piece of the lower joint
only was left, when feeling for the first time really
roused, he made one fierce rush and mad leap, and
the line not unreeling fast enough to suit him, he
disappeared with three flies, all my leader, and most
of my line. I do not advise any one to fish for salmon with an eight-ounce cedar trout-rod.

In ordinary trout-fishing, however, salmon do not
abound nor come unceremoniously devouring our
baits intended for their smaller brethren; nor are
even trout so extremely numerous but that, for a
long summer day's work, a light able rod will be infinitely preferable to a heavy one. A rod that
weighs fourteen ounces is heavy, and I have seen
persons with their hands or wrists dreadfully swollen after a single day's fishing, and have had such
persons assure me that their rods were as light as
they could be possibly made. Delicacy to me is the
first essential in trout-fishing, whether delicacy of
rod and tackle, or delicacy of handling and casting.
Catching a trout with a stick and a string is not
half the fun of catching a flounder, the latter being
much more difficult to lug out of water; and delicacy in trout-fishing will bring the best reward.

With a cedar rod you need use the wrist alone,
and that without much exertion; you can cover
great distances and still control the line, and you can
switch the fly under bushes and in difficult places,

better than with any other rod I ever used. It is quick, reliable, vigorous, and light, the slightest motion gives the tip the requisite spring, and it answers every effort of the hand instantly. It kills a fish powerfully and rapidly, and exposure to wet neither deadens nor weakens it. The ordinary hickory and ash-joint are much stronger, but are logy in their action and far heavier; joints of split cane or malacca are light, beautiful, and expensive, but are almost unattainable, and are, occasionally at least, deficient in power; and whalebone, for any part of the rod, is dull, heavy, inappropriate, and when water-soaked, utterly worthless. For these reasons and many others—these are enough, however—I prefer a cedar rod.

Many persons give the preference to a limber rod, one that bends in the middle, and they can, after infinite practice, cast well with it; in pleasant weather they can throw a light line, but when the storm lowers and the wind blows, or the current rages, or the cast is very long, or the bushes overhang, then good-bye to the gentleman with that most wretched of implements, a weak-backed limber rod. Give me no such inefficient deception to break my wrist, my heart, and my patience; as well tell me that whalebone has the vigor of a steel spring.

The joints of a rod are united in various ways; with the salmon-rod it is almost essential, and with all rods desirable, to use splices, but the custom is to indulge the laziness of ferrules. American ferrules fit accurately, and of course after the wood is swollen by exposure to rain, they will not come apart even if the joint-ends are all brass, a difficulty that can be obviated by rubbing them with mutton tallow, and loosening them every night, and we advise the same precaution in wet weather with the reel bands. In this connection it may be well to tell the reader how he can, with a little trouble, separate the ferrules, no matter how solid they may seem to be; in the first place heat them moderately, and pour a little oil round the joint; then take two stout pieces of string, or better, braid, about a foot long, and tying the ends of each together, wrap one close above and the other below the joint in the contrary directions; then insert a stick in each loop, and turn one one way, and the other the opposite. If the bands slip, rub them with wax.

The English ferrules, not fitting so closely, are not liable to this objection; but, on the other hand, would come apart in use, to the intense disgust of the angler, were they not held together by a piece of silk, that, when they are set up, has to be wound round a loop of brass fastened upon each for the purpose. This silk must be cut every time the rod is taken apart, and occasions much trouble. The Irish use a screw-joint, which is firm and not liable to bind; but it is difficult to fit, easy to break, and, in the woods, impossible to replace. Among these plans the simple socket has obtained the preference, and probably is entitled to the distinction.

It is doubtless useless for me at this day to tell any

intelligent sportsman that the butt of a fly-rod must never be hollow; its solidity is necessary to a proper balance; but where the fishing is merely to be done along the streams, a spear-head that can be screwed into the end will add little to the weight, and prove useful driven into the ground to hold the rod, while the fisherman changes his flies or frees them from a weed or bush. On a trout-rod there should be no reel-bands, but a gutta-percha ring, or a leather strap and buckle, will retain the reel firmly, and enable the angler to change its position at his pleasure, and by altering the balance, rest his wrist. These seem trivial matters, but mole-hills are mountains if they rest upon a sore spot. On a salmon-rod the reel-bands should be strong, and about a foot from the end.

There should be rings or guides enough on a fly-rod to bring the strain evenly throughout, and if one is destroyed, it should be replaced at once, or a liability to break will result. If rings are used, they and the brass top should be large and fastened on with a whipping of silk, that adds much strength to the wood. Where a spliced rod is used, it is well to have a small ring of brass, somewhat similar to the reel-band on each joint, under which the end of the splice can be slipped before fastening it.

For salmon and trout-fishing, the reel had better be a simple, large barrelled click-reel, as the music of the line, unwinding to the rush of these splendid fish, while it indicates the rate of its diminution, is to the angler what the clarion is to the warrior, or the hound's bay to the deer hunter; but a multiplier, made as they are only made in this country, working with the beauty and accuracy of clock-work, is by no means inadmissible. A drag must be used with the multiplier, but a stop never; the latter is utterly useless, and by slipping unexpectedly, may destroy your tackle. The reel must be manufactured with the greatest care and of the best workmanship; no implement is so worthless if poor, and none will better repay the sportsman if perfect. In salmon-fishing, it is only in desperate straits that any effort is made to check the fish; he is ordinarily too violent to submit to such treatment; otherwise, as the single-barrelled reel revolves toward you, it could not be used, as it cannot in bass-fishing.

A multiplier should have steel pins, which require care and frequent oiling; the same reel may be used for bass, and, if armed with a drag as above stated, in case of necessity in salmon-fishing. For both salmon and bass it should be of the largest size, and may be painted black to preserve it from rust, and to avoid alarming the fish. The line will occasionally catch round the handle, to prevent which, the latter is sometimes constructed of a button fitting in a plate.

All reels must be oiled occasionally. On one occasion I proved this to my satisfaction in a very unsatisfactory way.

The weather had been hot and dry; the water had fallen and become transparent as crystal; the fish were shy and cautious. After exhausting my ingenuity in selecting new flies to suit their capricious tastes, I had settled upon one of bright yellow, which, if the gentlemen did not wish to eat, they did seem to enjoy inspecting; they rose to it freely, and after I had tried in vain to strike them, curiosity induced me to keep count of their number.

Fourteen times had they risen and disappeared uninjured; fourteen times had my nerves tingled, and my blood started; fourteen times had sudden hope turned to bitter disappointment, till anticipation settled down into dull despair. Only those who have themselves had such painful experiences can appreciate my feelings; the continual tantalizing approximation to success, to be followed by agonizing failure; the renewed hope that the next rise would result in the capture of a fish ever to remain unfulfilled; the desperate effort to strike quicker or to cast more attractively; all these and many other feelings swarmed through my heart, as fish after fish approached his fate, and invariably escaped.

They seemed to be feeding, as it is called, and when the fly passed they rose, and turning over like a porpoise chasing mossbunkers, seemed to take it in their mouths. They did not spring out of water in the gaiety of reckless play, but acted as they would have done if swallowing the natural insect. Not that it is certain that salmon feed on flies; but while they can rarely be taken while playing, they often can be when acting in a manner resembling feeding.

My patience not exhausted, for it never is while fish will rise, I directed the canoe to be dropped towards the lower end of the fishing-ground, and stepped from it to a rock in the stream, and then casting the farthest and lightest possible, was rewarded. A magnificent fish rose; was secured by a quick turn of the butt, and stung by the unexpected pain, fled down the current. Away he went, on without a pause, the reel hissing, the line unwinding, and darting into the water, till having exhausted seventy-five yards of line, and being partially turned by its weight and the resistance of the click, he stopped with a heavy surge, and heading back, approached as fast as he had fled. Instantly and instinctively my hand fell upon the handle of the reel; it would not turn, no effort could budge it; conceive my feelings now, if mortal man can conceive them. The fish coming towards us, the line lying in a long heavy bag behind him, threatening to sink and catch round some rock, or by its slacking up release the hook; I jerked in the line, thinking a grain of sand might have penetrated between the plates, and tried the handle first one way, then the other, in vain.

This all passed with the speed of thought, but the fish was approaching as quickly; there was nothing left but calling one of my men to tell him to take in the line, hand over hand, and holding it in a loose coil, be prepared to pay it out on the next rush. Then thinking that the plates must be bent, I took from my pocket a screw-driver that I always carried, and unloosened every screw. There I stood, grasping in one hand the rod, while the tip bent to the motions of the fish, with the other working away at the reel; beside me my best man, slowly drawing in or paying out the line as need must; both of us eager, anxious, and startled at this new mode of killing salmon; the fish, vigorous as ever, making continual and sustained rushes, but fortunately none as extended as his first.

I had freed every screw in the reel, but without any result; it was as immovable as ever; there was no resource but to do the best we could, in our original mode of proceeding, under the circumstances. Never before had a fish-proved himself stronger or braver; for a good half hour he kept us on the stretch, and then sulked. Stationing himself in the edge of the current, he held his own doggedly; fifteen minutes of such behavior exhausted our patience. If I tried to lead him towards the shore, he took advantage of the eddy to resist; if to turn him the other way, he braced himself against the current; a severe strain, however, brought him to the surface, and revealed the fact that he was not sulking at the bottom, but resolutely swimming, head up stream, in the current.

Not a little surprised, we tossed in a pebble, then a stone, at last a rock, when, indignant, he fled down stream; fifteen minutes more of exciting contest, several rushes when he was on the point of being captured, resulted at last in bringing him flouncing on the gaff out of water. He only weighed fifteen pounds, but had been hooked foul, the point having penetrated at the hard bone near the eye.

I then sat down deliberately to discover what had

happened to my reel; it seemed to be in perfect order, but would not move; I tried to drive the shaft out of its bearing with the mallet—a heavy club of wood used to kill the fish after they are gaffed, but only after a good hour's work did I succeed in separating it, and found that for want of oil the two surfaces had become almost solid. They were as bright as burnished gold, and had evidently been heated by the first desperate rush of the fish; after being touched with a drop of oil and replaced, they worked beautifully.

It is curious to note how, in salmon-fishing, accidents will happen when the fish is on the hook; if the line is weakened, or the leader fretted, or the rod strained, the weight and power of the fish expose the weakness; if anything is aught but perfect, it gives way at that critical moment. In trout-fishing you are apt to discover the defects in time, and in bass-fishing the tackle is coarse and strong; but in salmon-fishing you first learn their presence by their parting. Never use a doubtful strand of gut, or a second-quality hook; never tie a knot without thoroughly testing it, and never use a leader that is in the least worn.

The best line by far, for both salmon and trout-fishing, is the braided silk covered with a water-proof preparation, and tapered to the fineness of the gut-leader. If this can be obtained no other should be thought of, but if it cannot, the others are about on a disgraceful par of mediocrity; the one that is usually praised, that of silk and horse-hair mixed, being, if possible, the worst, for while it has the weakness of the horse-hair, and water-soaking capacity of the silk, it has a difficulty especially its own, arising from the protrusion of short ends of hair that have broken or rotted off, and which are continually catching the rings or guides. The common silk line may be coated with raw linseed oil by stretching it in a garret or some place shaded from the sun, and rubbing it with a cloth soaked in the oil; several coats must be applied, allowing each to dry before a renewal, and care must be taken to avoid exposure to the sun's rays, which will rot the line. If thoroughly coated it will answer nearly as well as if prepared in a more scientific manner.

The elegance, ease, and delicacy of casting depend much upon the proportions of the leader or casting-line, its length, taper, and adaptation to the line and rod; if these are not accurately ascertained and complied with—and they can only be determined by actual experience with each rod and line—the execution will be faulty. Consequently no absolute rule can be given, but the length and taper must depend upon circumstances. The strands of gut are selected, the clearest, roundest, and hardest being the best, and having been assorted according to size, are tied together with the double-water knot for salmon-fishing, and with either the same or the single-water knot for trout. If it is desired to fasten the droppers between the knots, the latter must be used, and the gut must be well soaked in warm water before it is tied. Leaders thus prepared and suited accurately to the line and rod, will be found cheaper and more satisfactory than those usually sold in the shops, and may be tapered to any degree of fineness.

The fly-book in which the sportsman collects his treasures—the fairy imitations of the tiny nymphs of the waterside—and which is the source of so much delight in inspecting, replenishing, and arranging during the season that the trout are safe from honorable pursuit, is at present one of the most ungainly and inconvenient things that he uses. It is either of mammoth size and filled with flannel leaves in which the moth revel, but in which the hooks will not stick, or it is so ingeniously arranged that the flies on one page entangle themselves in a remarkably complicated manner with those on the other, and whenever the book is opened do their best to tumble out and carry with them such leaders as may be within reach of their obstinate barbs. It has places for articles that are not wanted, and none for those that are; the disgorger, an instrument about as useful to the angler as a jack-plane, is always present, while a piece of India-rubber to straighten gut, or even silk and wax, is never to be found. The pockets and slips are so arranged that the flies cannot be got at without much difficulty, or else fall out with perfect ease, and are invariably, when released, found with the gut so curled up that it cannot be straightened for some time. In fact, the present style of fly-book is a disgusting monstrosity. The true plan is to so arrange the pockets that those of one page will come opposite the hooks on the other in such manner that there can be no entanglement; of course the snells of the stretchers cannot be kept straightened, but the droppers, having shorter snells, may be secured under strips of paper, and left at full length, the alternate flies being at each extremity of the leaf; and on the adjoining leaf in the pockets may be similar flies dressed for stretchers. Or the droppers, all having the gut tied, of the same length by measurement, over two pins stuck into the table, may be secured on both sides of a separate sheet of pasteboard upon hooks and eyes, the fly-hook being fastened into the eye and the loop upon the hook. The latter is attached to a short piece of elastic, and will hold the gut straight and safe. The boards thus prepared are carried in long pockets between the leaves. The book, when filled and ready for use, should not be too large to be carried in the breast pocket, should be composed of stout parchment or ass skin that will resist the effect of dampness, covered with leather or morocco, and closed with a neat clasp.

The best implements will not make an angler, nor enable him, without skill that can only be obtained by patience and perseverance, to perform his duty creditably at the river-side. Especially must he learn to cast his flies far, lightly, and accurately, for of all the angler's qualifications this art is the most necessary. To do this every writer on fishing has given particular directions, but in reality no plan or formula can be made that is not subject to great modifications; the following, probably, is as nearly correct as any: After the line is lifted from the water, which is done with a quick upward motion of the wrist, the forearm is slowly and steadily raised until the line has described the necessary curve and is extended almost directly behind the angler, when a fresh impulse from the wrist changes the direction to a forward one, the arm following the motion until the line has nearly reached its limit, when it is checked by an almost imperceptible motion of the wrist, and the flies are made to drop on the water gently and quivering with almost the tremor of life. This is the rule when the cast is down wind and unobstructed, and the breeze light and equable, but in practice each cast must be adjusted to the peculiar circumstances under which it is made; the force that will drive out the line in a heavy breeze will not be vigorous enough if it dies down at the next cast, and the line must be stopped short or it will not extend itself; on the other hand, if the wind suddenly increases to a gusty flaw, the flies will be driven into the water with a splash, unless the arm is extended to exhaust the additional force. If the cast is across a strong wind, the line is lifted against it and makes almost a complete circle, and if well managed can be made to so resist it that, in the roughest weather, it will go out its full length and fall with beautiful delicacy. In a hard blow the difficulty will be in raising the line, and at times it will not be found necessary to lift the flies entirely from the water before casting, as the wind, by its pressure on the bag of the line, will carry them out of itself. In fishing a stream there is much to be learned in the art of jerking the flies under the bushes, and tossing the back line directly upwards to avoid entanglements, instead of behind the angler; proficiency should be obtained with the left hand as well as with the right, and in right and left casts, that is to say, where the line is raised on either side and the flies brought over either shoulder. This last point is essential if two anglers are to fish from the same boat, for each should invariably keep the tip of his rod over the shoulder opposite to his neighbor.

These observations are probably all that can be placed on paper with any advantage, for complete knowledge can only be obtained at the brook or pond under the guidance of those skilful teachers, patience and perseverance; and after the line has been neatly cast and the trout lured from his lair under the bank of the stream, or his mossy bed at the bottom of the pond, the art of striking him, that is, fixing the hook firmly in his mouth when he has grasped it, can only be acquired by actual experience. All written directions on this subject may be reduced to two—it is done with a motion of the wrist and as quickly as possible; and yet if this art is not mastered, the rest will be in vain.

There are few matters connected with fly-fishing that have been more discussed, and about which there has been more difference of opinion, than the length of line that can be cast with the ordinary

trout-rod. Assertions are common, and certificates
even have been given at public contests that competitors have cast one hundred feet of line, and many
persons, especially those not thoroughly initiated,
imagine that they can readily manage seventy, eighty,
or ninety. But this matter was brought to a defi-
nite issue at the convention of the Sportsmen's
Clubs of the State of New York, held in 1864, at
the City of New York, when a handsome prize was
offered for excellence in casting the fly, and rules
were carefully prepared to govern the trial. These
rules are given at length hereafter, and provide an
allowance, for length and weight of rod, and pre-
scribe certain distinctions as to whether the contest
is only as to distance, or as to delicacy and accuracy in addition. In the instance referred to, it was
determined that all these points were to be included.
No rod was admitted that weighed over one pound
or exceeded twelve feet and six inches in length ; a
gut-leader of not less than eight feet was required,
and to this three flies were to be attached. The
tackle and rods used by the competitors were, in
every instance, those that they were accustomed to
use in actual fishing, the lines being generally of
plaited silk, covered with the ordinary water-proof
preparation. The water was without a current, but
ruffled by the effects of a light breeze that died
away entirely ere the contest was over, and the stand
was a floating platform, level with the surface, and
upon which the waves occasionally washed so as to
wet the feet of the contestants. The distance was

measured along the water by a rope stretched taut
and marked at every foot of its length with buoys;
parallel with this, and close to it, a staging was
erected, on which the spectators could stand and
observe accurately the quality of every cast. The
contestants were required to use both hands, and
were restricted to five minutes' time. The judges
were three of the most experienced fishermen of the
State, one of whom is celebrated for his proficiency
in, and devotion to casting the fly.

It will be observed that several customary advantages were lost by this disposition, or brought to an
equality; there was no elevation above the water,
which is always difficult to measure, and which, of
course, adds immensely to the distance that can be
covered; there was little or no wind to add to the
forward motion of the line, and no current to
straighten it out, or assist, by a slight resistance to
the rod, in recovering it, which, after all, is the main
difficulty, as the line that can be lifted and extended
behind the fisherman will readily reach its full
length in front of him; and the distance cast was
measured, not along the line, which will invariably
sag more or less, and may have its length considerably augmented by an irregularity in delivery, but
along the water. Moreover, the competitors were
required to make a neat as well as long cast, lest
they should be ruled out for want of delicacy, and
had to prove their thorough proficiency by dexterity
with the left hand.

The rods used were respectively of ash, with a

split bamboo tip; of cedar, with a lance-wood tip; and of split bamboo throughout; and were all of the best workmanship and perfect representatives of their kinds; the contestants were some of the best anglers of the State, and nothing occurred to mar the pleasure of the contest or to disparage the correctness of the award. The prize was won by the cedar rod, which was twelve feet three and one-half inches long, and weighed, with heavy mountings, fourteen ounces; and the greatest distance cast with the right hand was sixty-three feet, although the allowance carried the official return to sixty-eight feet; and with the left hand the absolute distance was fifty-seven feet. The author cannot help adding that the cedar rod was in his hands, and that the prize is now in his fire-proof safe, as he thinks that success at such a trial and against such competitors is legitimate ground for no little vanity.

It is reported that there was a contest of a similar nature in England; but while the length of rod was restricted to twelve feet, there was no allowance for weight. The contestants stood several feet above the level of the water, and the distance reached was seventy-two feet. This, therefore, scarcely furnishes a ground for comparison, as a rod may be made so heavy at the top and limber in the middle as to cast a prodigious line, but which would be utterly unwieldy at the river side; and for every foot of elevation several feet of additional length are gained. In public trials attention must be paid to these particulars, or they will furnish no satisfactory test.

The writer once cast seventy-two feet with the same cedar rod that won the prize; but this, although without the assistance of any wind, was done from a slight elevation with the aid of a current, and was measured by the length of line. It is undoubted, moreover, that sixty-three feet is not the limit that can be attained where no attention is paid to delicacy in delivering the flies, or where but one fly or none whatever is used. The line can be cast considerably farther without a fly attached than with it, and the length and taper of casting-line should accord exactly with the weight and taper of line. This has to be regulated in a measure by practice, and should be carefully determined before a public trial is undertaken.

The author of the American Angler's Book recommends that the largest fly should be used as the stretcher. . This is all wrong, and no one that does so will ever deliver his flies far and neatly. It is contrary to the principle of tapering the line, and has no advantage whatever to recommend it. The largest fly should be the upper dropper or bob, and the next in size the second dropper, while the stretcher should be the smallest. Then not only will the taper be maintained, but if a trout rises at the droppers there will be more probability of striking him. One of the contestants at the trial above mentioned delivered his line so delicately that the flies often could not be seen to strike the water or make the least disturbance on its surface, although the spectators were close to the spot where they fell.

He was on a previous occasion ruled out of a contest because the judges could not see where his flies alighted. He is especially careful to maintain the true taper of line, casting-line, and flies, and would scout the idea of using a cast with its largest fly at the stretcher. This is as gross a heresy as putting a shot in the fly-hook, which, while it may tend to break the rod, instead of increasing will diminish the distance reached.

The author of the work referred to, although doubtless a hearty participant in the angler's pleasures, and fond of the free life in the wild woods by the side of the secluded stream, shows, by his preference for common flies and coarse tackle, that he does not appreciate the higher development of his art in its purity; content rather to fill his basket with a stout hackle from the well-stocked brook of the rarely visited forest, than to tempt the dainty trout with finer imitations from the well-fished pond of the cultivated country. Not only are large flies, especially at the stretcher, difficult to cast, but the hackles which he especially recommends are, from the resistance to the air offered by their numerous bristles, by far the most difficult. It is almost impossible with a light rod to cast a large hackle delicately to a distance; and when three are used, it is entirely so. In clear pools such an apparition would frighten the trout from their "feed" for a week. But in a boisterous, roaring, foaming mountain cataract, where the fish cannot see the fisherman at all, and find difficulty in seeing

their prey, hackles and palmers are perfection.

The foregoing match was governed by the following rules, which have been permanently adopted by
the New York Sportsmen's Club, but the allowance
of time is not sufficient where delicacy and distance
both are to be determined ; and the better plan would
be to allow each contestant first to extend his line
as far as he can, and then to restrict him to five
minutes as to the other matters at issue.

RULES OF THE NEW YORK SPORTSMEN'S CLUB, FOR CONTESTS IN FLY-CASTING.

No *Rod* shall be allowed over twelve feet six
inches in length, nor more than one pound in weight,
and it shall be used with a single hand.

A practicable *Line* and *Click-Reel* shall be attached to the rod.

One *Stretcher Fly* must be used, and a *Casting-
Line* or *Leader*, of single gut, of not less than six
feet in length.

Additional *Flies* may be added in the discretion
of the contestants.

No attached *weight* of any kind on the line or fly
shall be permitted.

Allowance of distance shall be made according to
the length and weight of each rod of five feet for
every foot of *length* and two feet for every ounce of
weight, and at that rate for a part of a foot or ounce,
deducting for a hollow butt or the omission of the
customary mountings.

Each *contestant* shall be allowed *five minutes* for casting, and in case of accident, such as the parting of the fly, or entangling of the line, the referee may once allow *additional time*, in his discretion.

No cast shall be valid unless the line be *retrieved*.

The character of the contest, whether as to *distance, accuracy,* or *delicacy,* shall be stated at the time of making the terms, and, if not so stated, shall be only as to the distance, which, if practicable, shall be measured along the water.

In case delicacy and accuracy are to be considered, the casting shall be done with each hand, across, against, and with the wind, in over and under casts, and not less than three flies must be used on a leader of at least eight feet in length.

Salmon Fly-Casting.—The above rules shall govern, unless it shall be distinctly agreed that the contest is to be with double-handed rods, in which case they shall be modified as follows:

The rods shall not be over twenty feet, and the casting-line or leader not less than ten feet in length.

Allowance of distance shall be made for length, but not for weight, and no more than one fly shall be used in any event.

In addition to the imitations of the natural fly, efforts have been continually made to use artificial representations of the other food and baits for fish; exact and beautiful copies of grasshoppers and frogs have been constructed, and painted of the proper color, but either from the nature of the composition or some other cause, entirely in vain. Indeed it is doubtful whether any fish was ever captured with such delusions as grasshoppers, crickets, or frogs, and although they are still retained in the shops, they no longer find a place amid the angler's paraphernalia. Squid and spoons are usually supposed to imitate minnow, and have always been to a greater or less degree successful, but the imitation fish itself has, until late years, invariably proved a failure. With the discovery of the proper preparation of gutta-percha, and its application to the innumerable purposes for which it is now employed, came the suggestion that it might in various ways serve the angler; as wading-boots and water-proof clothing, of course, but also for bait-boxes, rods, and finally minnows. A little fish made of this material is not only a faultless imitation of the original, and is even curved in a way to produce the most perfect spin, but being soft to the teeth, seems absolutely to convince the trout in spite of their palates that it is wholesome and appropriate food. This imitation is used with satisfactory results, not only for trout to which it is peculiarly adapted, but also for snapping mackerel and lake-trout; it is so admirably prepared that the eye cannot detect the deception, and it has about the same consistency as fish itself. The back is a delicate mottled green, changing to yellow on the sides, where there are a few vermilion spots, while the lower part is brilliant and sparkling with some preparation of quicksilver. There is a gang of three

12*

hooks near the head and another at the tail, which is of tin, and the whole is attached to double gut. A modification of the same article is made by fastening two tin flanges at the head of the same minnow and leaving the body straight, but by the change more is added to the weight than to its effectiveness.

This invention is extremely light, being hollow, can be cast even with the fly-rod, and has been known to do great execution. In its present perfected form, it is a foreign production, but the original discovery was American. It is especially successful with lake-trout, even more so than with brook-trout, but is too delicate to trust in the hungry jaws of a savage pickerel. When the snapping mackerel first appear, and before their increasing appetites have made them as ravenous as they subsequently become, and when they will not condescend to the leaden squid, they will readily take this gutta-percha artificial minnow. One of its great recommendations is its lightness; no imitation bait that falls with a loud splash into the water can do other than terrify the timid trout; and to make casting a pleasure, the rod must be delicate, which cannot be if the bait is heavy. The squid is usually supposed to be the original imitation of a minnow, and has held a prominent place among the angler's delusions for many years; in bass-fishing, in trolling for blue-fish, and even for lake trout, it is worthy of all praise. For bass, it is true, the natural squid is far more tempting, but this queer monstrosity is difficult to obtain, and its substitute has often captured enormous fish; for blue-

fish no other bait is ordinarily used, and for lake-
trout the ivory squid can hardly be surpassed. The
ordinary kinds are of lead, pewter, bone, which
are often hollow, and admit the insertion of a large
hook; and of pearl, the latter in its most killing
shape having flanges and spinning like the minnow.
For blue-fish and their young—the snapping mackerel, lead is the favorite, while for lake-trout and
pickerel, ivory is preferable, although this rule is
not invariable; and on dark days the light-colored
material will be occasionally preferred by all these
varieties.

As the trolling-spoons resemble no known crea-
ture, they also are supposed to be intended and
accepted for the minnow, although it is difficult to
conceive why fish with their sharp sight, that can
distinguish an almost microscopic midge upon the
surface of the water twenty feet above their heads,
should mistake a piece of revolving tin for a living
fish. The first of these contrivances were manufactured and named from the bowl of a pewter spoon,
the handle being broken off and holes drilled in each
end, so that the line and hooks could be attached;
this bait was found to revolve and glitter in the
water in an attractive way. It is now almost super-
seded by other modifications; but still, when made
of bright tin and painted of a dark color on the con-
vex side, and rather more elongated than the ordinary pattern, it is successful with lake-trout and
Mackinaw salmon. The first alteration in shape was
by fitting two flanges or wings on a long, hollow

body, upon the principle of a screw, and named after Archimedes, by which a rapid revolution was produced; but although this invention seemed to man nearly perfect, it did not satisfy the fish; for a very small spoon it will answer, but when larger is not so attractive as other kinds. Several alterations and combinations of these two plans were produced from time to time; they proved to be merely changes and not improvements, until an invention was made that is usually called Buel's Patent Spoon—although it has been said that his patent only covers the application of three hooks instead of two, and that the invention has long been in use among the pickerel fishermen of the St. Lawrence. The blacksmiths on the banks of that river certainly manufacture them unrestrainedly of such material as they prefer, but only use two hooks; and this would not probably be permitted if the patent was broad enough to prevent it.

Be that as it may, however, it is known as Buel's Spoon; it is made by fastening two or three hooks back to back, and attaching a piece of tin nearly elliptical in shape, so that it can revolve freely round a collar at the shank. This is its simplest form, and the one preferred for mascallonge, for which two strong thick hooks are used, firmly soldered together; and for pickerel, black-bass, and lake-trout, it is safer to have the hooks either soldered into one piece or attached by wire, as the fierce struggles and sharp teeth of these species will soon destroy thread or silk. The tin is painted of various colors, or even replaced with brass, and should be kept well burnished on the bright side. Feathers of gaudy colors, such as ibis, golden pheasant neck, mallard, and wood-duck, interspersed with plain white, are often fastened along the shank; spoons thus prepared are favorites of the black-bass, but have no advantage for mascallonge over the bare hooks; they are also used successfully for trout, especially those captured in salt water, and the feathers as well as the coloring of the tin may be adapted to the state of the weather. On clear, sunshiny days dull colors are preferable, as with artificial flies; and in dark or rainy weather the lightest colors answer best. Three additional hooks are sometimes added, and allowed to dangle loosely below the others; although these occasionally capture a fish that has missed striking the spoon fairly, they are more frequently bitten off; they are really no advantage, and if once imbedded in the bristling jaws of a gasping pickerel, their extraction is both difficult and dangerous.

Of the different varieties of artificial bait, not of course including the artificial fly, the most general and successful is Buel's Spoon; it is taken by all the pickerel, from the monstrous mascallonge to the tiny native of Long Island; by the trout of lake or brook; by the black-bass of the North and South, and by the young blue-fish of the salt water; it is generally a greater favorite than the artificial, and sometimes even than the natural bait; with black-bass it has no competitor but the fly, and with sea-trout it occasionally surpasses the artificial fly itself. Its irregularity of motion, consequent upon the mode of revolution, seems to be its charm; and although it does not spin as well as the Archimedes, it is infinitely more killing. It has in open water almost supplanted the use of bait for pickerel and mascallonge, and it has been used to a murderous extent by greedy fishermen in trolling the waters of Moosehead Lake for trout.

COOKERY FOR SPORTSMEN.

AMONG all the arts and sciences that improve, elevate, or embellish society, or that contribute to the pleasure and comfort of mankind, the one that is the most necessary to health and happiness, has produced the fewest great geniuses, and is the least under stood, is cookery. Amid the thousands of men and women who pretend to a knowledge of its mysteries, how difficult is it among the former, and how impossible among the latter, to find a good cook—one who is devoted heart and soul to the intricate science, who passes days in pondering and nights in dreaming of these delicate combinations that constitute pure and refined taste!

The world has produced in hundreds painters that delight the eye, composers that enrapture the ear, scholars that convince the intellect, poets that touch the heart; but of culinary artists that enchant the stomach, the truly great may be counted on the fingers. In ancient times more attention was paid to gastrology, but the degraded taste that could employ an emetic to enable the repetition of indulgence, and the limited resources of restricted national intercourse, have left us little of value to be gleaned for the experience of antiquity. The great masters of the kitchen of those times have passed away into oblivion, or have left only a few crude dishes, remarkable more for their extravagance than their excellence. It was a deficiency of knowledge and high art that drove the gourmands of early days to peacocks' brains, nightingales' tongues, and dissolved jewels.

The middle ages have left us some right royal dishes; the boar's head, the roasted ox, the black pudding, mince-pies, the plum-pudding; remarkable, however, more for their substantial character that satisfied a vigorous appetite, than for delicacy that would gratify an educated taste. During this period, however, many drinks attained a perfection that has never been improved on, and those delicious combinations that were called cardinal, bishop, punch, and the hearty sack, are almost as well known and as great favorites now as then. There is nothing to be drawn from the dark ages in the least elevating to the science of gastronomy, and we must look to modern times, and mainly to the French nation, for our highest authorities and truest instruction.

Catherine de Medicis introduced the art of cookery into France, and liqueurs were invented during the reign of Louis XIV., since which time the revered names of Vatel, Soyer, Ude, Kitchiner, Bechamel, and Carmel have become household words throughout Christendom; their skill has shed a benign influence over mankind, has restored invalids to health, and brought peace to families; they are quoted and looked upon with deep respect by all.

Coarse minds, to whom the allurements of gastronomy are incomprehensible, consider cooking vulgar; while a few pitiable individuals are created without the sense to distinguish the tasty from the tasteless, as there are persons without an eye for the beauties of nature or an ear for the harmony of sounds. These unfortunates deserve our sympathy; but for the individual who affects to despise the pleasures of the table, as loftily placing himself above what he terms grovelling appetites, nothing is appropriate but contempt. Who would believe or respect the man who claimed that his inability to distinguish green from red was a credit to him? Or could tolerate one who was filled with ostentatious pride because, by a wretched malformation, he could not tell Old Hundred from *Casta Diva?*

The sense of taste is as noble, and as capable of education and improvement, as the art of the painter or the musician. The stomach being the governor, master, and director of the body, when it is pleased the intellect works with force, the eye and ear are in full play, and the nerves and muscles tingle with animation; when it is sick or exhausted the eye grows dull, the intellect feeble, the ear inaccurate, and the whole body drooping and spiritless. It has its ramifications in every part of the system, and controls as inferiors the other organs. An ill-cooked dinner has lost many a battle, ruined many an individual, and disgraced many a genius; it is said that an indigestible *ragout* cost Napoleon his crown.

Life is dear to all, and yet persons are continually

committing a disagreeable and prolonged suicide, accompanied with painful indigestions and untold sufferings, by attempting to despise the rules that the imperative stomach has laid down. Under certain well-known chemical laws, food is rendered both digestible and palatable by special modes of preparation, and indigestible and unpalatable by other modes. The same piece of meat that, fried, will resemble shoe-leather, and afford neither pleasure nor sustenance, if nicely broiled would prove agreeable to the palate and wholesome to the body.

Our country is overflowing with abundance of the raw material from which good dinners are made; but we are absolutely without cooks, and the average American life is shortened one-tenth by the miserable ignorance of the rules of cookery that pervades all classes. The farmer bolts his heavy griddle cakes and tasteless fried meats; while the wealthy citizen devours rich gravies and ill-prepared compounds. The former loses his teeth, the latter incurs the thousand horrors of dyspepsia, and both shorten their lives.

But to rise above the unimportant consideration of mere life, which is held in our land at its true value, and regarding cookery from a loftier point of view, is there not something noble in the art that moulds together the various subjects of taste, and builds up an exquisite, soul-thrilling composition? Is not that man worthy of our deepest admiration, who, not only from the wealth of materials prepares the perfection of luxury, but when reduced to the sim-

plest articles, still manages to gratify the most delicate and exacting of our organs? Who has not felt
his heart expand as he surveyed a royal feast; his
affections become purified, his feelings elevated, as
dish followed dish, and each proved itself worthy of
the other; and at last has not taken a gentler view
of human kind when contentment filled his soul?
A good dinner encourages generosity, begets sympathy, increases geniality, while it strengthens the
intellect and the nerves; a bad dinner produces ill-
nature, leads to discontent and quarrelling, dulls
the mind, and injures the body. The former aids
Christianity and promotes virtue; the latter is the
bold accomplice of vice and crime; evil humors cannot exist in the body without spreading to the mind,
and vices in the former create vices in the latter.
Controlled by that complacency which is the stomach's return for kind treatment, the evil passions
sleep, and fading gradually, lose half their strength;
whereas, if aggravated by perpetual dissatisfaction
and uneasiness, they become daily more violent, till
they disdain command and burst forth in unrestrained fury. So that the soul, even, may be endangered
by bad cookery. The civilization and power of nations advance in proportion to their improvement
in their cuisine, and the reformation is said to be
due to the strong Teutonic impatience of fast days.
A coarse taste in eating is as sure an indication of
coarseness in mind and habits, as delicacy of taste is
of delicacy and refinement in other particulars. As
the more vulgar desires are controlled by the

higher impulses of the mind, and clean hands are
often the index of a clean heart, so purity of appetite usually accompanies purity of soul. Nothing
condemns the vulgar man more quickly than the
nature of his appetite, and his mode of gratifying it;
driven on like the beasts by hunger, he thinks only
of the readiest and quickest mode of satisfying the
unpleasant craving, and never dreams there can be
anything intellectual in a dinner. The Americans,
as a nation, are ignorant of the first principles of
dining; in private, they ruin their digestions; in public, they disgust their fellows. With that practical
turn for which they are famous as a body, they devote themselves to what is profitable; and the arts
of sculpture, painting, and gastronomy are just beginning to be appreciated.

Those huge dishes that delight hungry, vulgar
John Bull, such as roast beef, boiled mutton, and
the like, still meet with the approbation of the active
American; and while our women, with their natural
elegance, draw their fashions from France, our mat-
ter-of-fact men imitate the rude cookery of England.
It is a melancholy truth that there is no place in
America where a dinner can be obtained; feed-
ing-places, miscalled restaurants after those priceless
legacies of the French revolution, are innumerable;
but even the famous Delmonico fails to appreciate
that wonderful production, the pride of our land—
none of the miserable little coppery European
abominations, but the great American oyster—does
not understand it, and never rises to a proper com-

prehension of its capabilities, and consequently never serves a perfect dinner.

So must it be while ignorant Irish cooks—whose only claim to the title consists in having spoiled thousands of potatoes, in having rarely seen, and never cooked, a piece of meat, and only dreamed of coffee—possess our kitchens and rule the roast; and as it is impossible for the master of the house, and would be unladylike in the mistress, to superintend the dinner, the only spot for truly scientific cookery is in the woods. There, under the blue vault of heaven, where the shade of some friendly tree tempers the combined heat of sun and fire, accompanied only by the interested and appreciative guides, with the hot wood fire rapidly forming its pile of glowing coals, can the contemplative man, tempted by appetite and opportunity, devote himself to the higher branches of epicurism. Not that the materials are plentiful, rich, or costly, but working up from the very plainness of his fare a more gratifying compound. With that bed of coals suggesting broiling, and that dancing, smokeless blaze inviting roasting, no intelligent being would think of frying meat.

Under such circumstances, the larder being necessarily limited, and repetition threatening to breed disgust, ingenuity is sharpened and exercised to produce variety; an accurate knowledge of the power of different sauces is obtained, and new modes of dressing simple articles invented. It is to lead the mind of the reader in this direction, and not with the hope of instructing Irish cooks, or educating

American taste, that this short article on cookery is
written; and if the life in the woods, or on the water,
of our sportsmen shall be in a degree improved by
the effort, the main object will be attained.

The materials generally at the disposal of the
hunter or fisherman on the coast and in the woods
consist of fish, oysters, clams, ducks, game birds,
and venison; while he will carry of necessity pork,
ship-biscuit, salt, and pepper, and, if possible, eggs,
flour, sauces, Indian-meal, and as many of the minor
aids of a good *cuisine* as his means of transportation
will admit.

No attempt will be made to confuse the reader
with complicated directions for the construction of
highly seasoned and strangely named French dishes,
but the simplest and readiest mode of cooking each
article will be given, with instructions in varying
the effect. If the enthusiasm inherent in the subject shall occasionally carry the writer away and
lead him to indulge in what the reader—living on
hard tack and salt pork—may regard as vain ima-
ginings, the weakness of man in the contemplation
of so vast a subject must be the excuse; and the
disciple need undertake nothing for which he has
not the materials.

One of the great deficiencies, although partially
supplied by the solidified article, is milk, which cannot be kept in its natural state, and is badly repre-
sented by its substitute. Generally, however, water
will answer in its stead, and for gravies or thickening for stews, a little flour mixed with a lump of

butter, and dissolved in a cupful of tepid water, is an excellent equivalent.

Oyster Stew.

The American oyster, to the thoughtful mind, presents itself almost as an object of veneration, and would among barbarous nations have altars raised to its honor; to the practical mind it is a mine of luxury, a very Golconda of epicurean wealth; raw broiled, baked, roasted, fried, stewed, or scolloped, it is the tit-bit of perfection, and in every mode may be varied extensively; it takes all flavors, and is delicious without any; it is improved by all sauces, and needs none. It accords with every other dish, or makes a dinner alone. The subject has never been half explored, much less exhausted.

A stew may be made with crackers or flour, with celery, cheese, or milk, and with or without sauces; but in every instance the juice must be separated from the oysters and well cooked before the latter are added, or they will be over-done, shrivelled, and ruined. The simplest mode is to put some pepper, salt, and butter in the juice, boil it five minutes, add the oysters, and cook for one minute longer.

Or you may add to the juice crackers pounded fine and rolled in butter, and some celery chopped fine, or a little cheese and Worcestershire or Harvey sauce; or you may put a table-spoonful of flour and as much butter in a cup, and having rubbed them together and added a little of the warm juice, may mix this slowly with the rest. This must all be done before the oysters are added; and where flour is used, care must be taken to mix it first with a small quantity of fluid, or it will lump. A dry stew, which is preferred by many, is made by cooking the oysters, from which the liquor has been carefully strained, in butter, salt, pepper, and sauce.

FRIED OYSTERS.

Dry each oyster separately on a towel; dip them in the yolk of eggs beaten up, and then in pounded crackers that have been seasoned with salt and pepper; heat butter or pork drippings in the frying-pan, and cook the oysters over a slow fire, turning them frequently. Do not use too much butter or drippings, but add fresh as required, so as to leave the oysters dry when done. A clean tin pan is the best, and red pepper preferable to black. Lard is detestable for frying anything, and salad oil is perfection. If black pepper is ever used, it should be purchased whole and ground by hand, as the fine pepper is generally adulterated and flavorless.

ROASTED OYSTERS.

To roast an oyster, it is simply put on the fire till it opens, when the shell is forced off, and it is eaten from a hot, concave shell, in which butter has been melted with vinegar, salt, and pepper; or it may be taken out when half done, and cooked in a pan with its own liquor, salt, pepper, and a little butter.

BROILED OYSTERS

Are prepared as for frying, then dipped in melted

butter, placed in a double gridiron, and cooked over live coals.

SCOLLOPED OYSTERS

Are placed in a deep dish with butter and bread-crumbs, or pounded crackers well seasoned and baked.

CLAM-BAKE.

The only proper mode of baking clams was discovered by the aborigines, and was invariably practised by them on their yearly visitations to the sea; the clams are placed on a flat rock side by side, with their sharp edges down and the valves up, and when so arranged in sufficient numbers, are kept in their places by a surrounding circle of stones. A large fire is built over them and allowed to burn for about twenty minutes, when it is cleared away and the clams are extracted from the ashes, overflowing with juiciness and steaming with aroma. Burnt fingers and lips add to the pleasures of an Indian clam-bake. The best sauce is pepper-vinegar.

CLAM OR FISH-CHOWDER.

Pork, potatoes, butter, crackers, sauce, salt, pepper, vegetables, and meat, if any can be had, clams or fish, or both, are covered with water, placed in a close vessel, and stewed slowly till patience is exhausted, appetite insists upon indulgence, or the mess threatens to burn. The large articles are cut in pieces of an inch square or thereabouts, and may be highly seasoned.

13

STEWED CLAMS, OR CLAM SOUP.

Hard clams are not fit to eat, stew them as you will. Soft clams, after the tough parts are removed, are excellent stewed with a little butter, or butter rolled in flour, as directed for oysters; but being richer than oysters, they do not need so many additions. The soup is made by thinning the juice before it boils with milk, which will curdle if thrown into the boiling liquid. Hard clams make a good soup if they are cut fine and not eaten.

FRIED OR BROILED CLAMS.

Soft clams may be prepared as directed for oysters, the tough parts being first removed.

SCRAMBLED EGGS.

Eggs are broken one by one in a cup to make sure they are fresh, and then thrown into a pan with a lump of butter, some salt and pepper, and stirred carefully, so as not to break the yolks immediately, over a slow fire till the whole is almost hard. They had better be too soft than too firm.

POACHED EGGS

Are broken into a cup and poured one by one carefully into hot water, and when done are ladled out on a flat, broad stick or spoon, so as to let the water drain off.

FRIED EGGS.

Fried eggs are broken one at a time into a cup, and poured into hot grease.

Omitted

Omelette.

Eggs are broken into milk, thickened with a moderate quantity of flour, salt, and pepper, which is beaten up and fried with butter; parsley, ham, or bacon may be added, cut fine.

Smoked Beef

May, be fried in grease with a little pepper, or may be stewed in milk. A little flour rubbed with butter in a cup, and mixed with some of the warm gravy, may be added.

Boiled Fish.

There are two modes of boiling fish; one recommended by Sir Humphrey Davy, and the other by the great Soyer. By the former, the fish cut into pieces is thrown into boiling salt and water, one piece at a time, and the largest first; by the latter it is placed in cold water, heated slowly, and allowed to simmer by the fire. The former, in his *Salmonia*, page 120, quotes chemistry to show that by the excessive heat the curd is coagulated at once and preserved; the latter refers to his unequalled reputation. I have generally pursued the former course as the more rapid; the water must be allowed to recover its heat after each piece is thrown in, so that it may be always intensely hot; about fifteen minutes of hard boiling will be required, but the only reliable plan is to examine and try the fish with a fork from time to time, as it is ruined if cooked too long, and uneatable if not cooked enough.

In Soyer's receipt the fish is placed in cold water that contains a pound of salt to every six quarts, which is then heated to the boiling point and allowed to simmer for half an hour if the fish weighs four pounds, for three-quarters if it weighs eight pounds, and so on.

Of course, a fish must be scaled ere it is cooked, and should be cleaned, although if it is cooked whole and the party is hurried, the latter process may be omitted without injury; the entrails, however, are not to be eaten.

A little of the liquor in which the fish has been boiled, with Harvey or Anchovy sauce, or Chili vinegar, makes an excellent dressing; but the best sauce is obtained by dissolving a spoonful of flour, that has been thoroughly mixed with a lump of butter, in a little warm water, and boiling the whole for a few minutes. This may be prepared in any tin pot, and, cooked with chopped parsley, is the making of boiled fish.

FRIED FISH.

The fish, which should be small, after being cleaned and scaled, are dipped in water and then in Indian-meal, and fried, well seasoned with pepper, in the pan with pork drippings or butter. If the latter is used, salt must be added. Trout are excellent prepared in this manner.

BROILED FISH.

Fish for broiling may be larger than for frying;

they are scaled, split open down the back, and well seasoned. They are placed on the gridiron and approached for a few moments close to the fire, so as to sear the pores. They are then cooked more slowly and well basted with butter, unless a piece of thin pork is laid across them, the grease from which will answer the place of basting. A favorite way to cook a shad or blue-fish alongshore is to split him entirely in two, and tacking the halves, seasoned and buttered, to shingles, to roast them rapidly; each man eats from his own hot shingle.

BAKED FISH.

Small fish or pieces of fish, cleaned, scaled, and seasoned, may be rolled in oiled paper and baked in the ashes ; or a whole fish unscaled, but cleaned and wiped dry, may be rolled in damp leaves and buried deep in hot ashes. When it is done, the skin and scales will come off together.

STEWED FISH.

Cold fish may be cut up into small pieces, seasoned and stewed in water, with a little salt pork. If milk is substituted for water, the dish will be more palatable.

LOBSTERS

Must be boiled when alive till they turn red. For a dressing the yolk of a raw egg is beaten up, with a tea-cupful of salad oil poured in very slowly till it is firm ; a tea-spoonful of mustard, a little salt, pepper, and vinegar are added and beaten together, after which more oil may be added, if necessary. The meat is picked from the shell, cut up fine, and mixed with a few spoonfuls of vinegar; the dressing is then poured over it.

Or the dressing may be omitted, and the meat cut into pieces may be warmed up in milk and butter, with pepper and salt, and served hot.

Potatoes

Are usually boiled by being thrown, after they have been washed, into an iron pot filled with cold water and a little salt, placed on the fire till the water boils, and allowed to cook till they are done, which is ascertained by puncturing them with a fork. The water is then poured off, and they are allowed to steam near the fire for a few minutes.

When cold they may be cut up and fried in grease, or mashed and stewed in milk, or mixed with small pieces of salt pork or meat, and made into a species of hash; in either case they must be well seasoned, and are improved by the addition of onions.

The best way to fry them is to slit thin pieces from the raw potatoes, and letting them drop into cold water, leave them for a few minutes. When taken out and fried in butter, they will be crisp and fresh.

Potatoes are tender and mealy if simply baked in hot ashes, which can be done by burying them under the fire until they become soft.

BOILED MEATS.

Meats are placed in cold water with a little salt, and boiled slowly, the scum that rises being removed from time to time.

FRIED PORK OR BACON.

Pork is cut into thin slices and freshened by being heated in the frying-pan with a little water. It is fried without any addition whatever, and the grease fried out of it is saved for cooking other articles. It can be breaded by being dipped first in cold water, and then in crumbs or Indian-meal, and fried crisp.

The same directions apply to bacon, and both should be cut exceedingly thin.

STEWED, BAKED, AND BROILED MEATS.

Meat may be stewed, baked, and broiled, much as has been heretofore directed for fish. In stewing, the great point is to proceed slowly, and in broiling to close the pores by burning the outside slightly on the start; and the next point is to season sufficiently, as both pepper and salt lose their strength in the presence of heat.

SOUPS

Are made by boiling a fish or a piece of meat very slowly; if salt meat is used, it must have been boiled previously in a different water; remove the scum till no more rises, add any vegetables, and boil till done. Use a quart of water to every pound

of meat, and keep the pot well covered. Rapid
boiling throws off the volatile portions of the meat
in steam.

Roast Duck.

Dip a duck or other large bird, neither cleaned nor
picked, in water so as to wet the feathers, and throw
him on the fire or into the hot coals. When the
feathers are pretty well singed, he is done, and the
skin, feathers, and dirt may be peeled off together.
A duck needs little more than a thorough heating.
Small birds may be rolled in oiled paper and roasted
in the ashes, or a bird picked and cleaned may be
suspended by a string near the fire, and made to revolve by twisting it up occasionally.

Beans

Should be soaked over night, and then well boiled.

Rice.

A cupful of rice is thrown with a pinch of salt
into enough boiling water to cover it well, and boiled
for fifteen minutes. It must be soft, but the grains
should be separate. The water is poured off, and it
is dished up hot.

Gravy.

White gravy is made as already directed for fish.
For brown gravy, a little flour is heated in a frying-
pan, and stirred till it is brown. It can be kept in a
bottle, and is added in small quantities to thicken
the juice of meat or soups.

Tough Meat.

Scalding vinegar may be poured over tough meat, which is left to stand over night; next day the meat is to be cut into small pieces and stewed with seasoning, and a few slices of potatoe and carrot.

Vegetables

Must be placed in boiling water with a pinch of salt, and are done when they sink; they must be taken up immediately.

Water Souchy

Is made by stewing fish cut into small pieces with chopped parsley and onions, and some pepper and salt. It may be poured over toast and thickened with flour and butter.

Potted Fish.

Small fish, cleaned and seasoned, and placed with a little mace in a pot lined with paper, are covered with melted butter, pressed down, and baked four hours with a weight on them.

Boiled Salmon.

Bleed the fish the moment it is taken by cutting its gills, and across its sides, in a slanting direction at every two inches. Hold it by the tail for a few minutes in the stream, moving it so as to encourage the flow of blood. Put the pot, filled with cold spring water, on a brave fire, so that it may heat while you are cleaning and scaling the fish. Divide into slices through the backbone, where the slashes have already been made. When the water boils, add a large bowlful of salt, and when it has recovered its heat and is screeching hot, throw in the pieces of salmon, the largest first, allowing the water to recover its temperature after each. For fish under nine pounds, allow ten minutes, and one minute more for every additional pound. Serve with a little of the brine strengthened with anchovy sauce, or make a white gravy of flour and butter, as heretofore directed. Save the brine for future use.

Trout on First Principles.

Catch your trout, put a pinch of salt in his mouth, roll him up in a few folds of newspaper, dip the swaddled darling in the water, light a fire, and place him in the embers. When the paper chars, take him out and eat him at once, rejecting the entrails.

Kippered Salmon.

Divide the fish down the back and remove the bone; rub him with equal quantities of sugar and salt, and a little pepper; dry him in the sun or smoke. Cut into thin streaks, and broiled, he will be found good and appetizing.

Daniel Webster's Chowder.

Four table-spoonfuls of onions fried with pork.
One quart of boiled potatoes well mashed.
One and one-half pounds of ship-biscuit broken.
One tea-spoonful of thyme.

One tea-spoonful of summer savory.

One half bottle of mushroom catsup.

One bottle of port or claret.

One-half nutmeg grated.

A few cloves, mace, allspice, and slices of lemon, and some black pepper.

Six pounds of sea-bass or cod, cut in slices.

Twenty-five oysters.

The whole to be put in a pot, covered with an inch of water, cooked slowly and stirred gently.

LIVER.

Pieces of deer-liver may be impaled on a red cedar skewer, with a slice of pork on top, and set up round a fire, near enough to cook slowly; the pork will melt and baste the rest.

GRIDDLE CAKES

Are made by thickening flour with milk or water, and adding an egg or two, together with a pinch of salt. They are poured in ladlefuls on a hot griddle or frying-pan that has been well greased. Rice that has been boiled and left over, or corn-meal that has been scalded, may be mixed with the other articles, and makes rice or Indian cakes.

CORN BREAD.

Two cups of Indian meal and one cup of wheat flour are mixed with two tea-spoonfuls of cream of tartar, to which is added one pint of sour milk or of sweet milk in which one tea-spoonful of soda has

been dissolved, beaten up with two eggs. The whole
is to be baked one hour. Cream of tartar is always to
be mixed with the flour, and soda with the milk, so
that when these are subsequently brought in contact, gas is evolved and the bread is rendered light.

<h3 style="text-align:center">SCOTT'S CHOWDER.</h3>

The following recipe was furnished by Mr. Genio
C. Scott to the New York *Spirit of the Times*, and
is doubtless equal to the reputation of the author :—

"The old-fashioned iron pot is the best to make it
in, but in lieu of it a copper-bottomed saucepan, as
deep as it is wide, will answer. First take your fish
—almost any kind will answer—but cod and sea-bass
are the best ; clean and scale your fish, and cut them
into pieces two inches square ; parboil a few onions ;
peel a few potatoes and quarter them ; cut up some
salt pork into the thinnest possible slices, and cover
the bottom and sides of your pot with it to prevent
your chowder from burning ; place upon the pork a
layer of fish, and season it with salt and a little black
pepper. (Since I read 'My Peninsular Medal,' I
have been very chary of black pepper, for that
authority states that it inflames the stomach without
stimulating it, while the cayenne pepper stimulates
without inflaming ; but a dash of black pepper is
useful for its fragrance.) Next, a layer of the par-
boiled onions quartered ; next, a layer of potatoes,
and season the layers ; next, a layer of ripe tomatoes
sliced and seasoned (tomato requires more salt than
other vegetables) ; next, a layer of cracked sea-bis-

cuit; next, a layer of fish ; then sprinkle this layer with infinitesimal pieces of salt pork, but sparingly ; then your layers of onions, potatoes, tomatoes, and sea-biscuit, with proper seasonings of each layer; pour water enough to cover the contents of the pot, but no more; cover the pot and place it on a slow fire where it will simmer or boil slowly for an hour and a half; a half hour before dishing the chowder, pour upon it a bottle of Burgundy or claret.

"In seasoning the different layers of the chowder, tomato catsup will answer where ripe tomatoes cannot be had. Sauces are also introduced sometimes, and in case the party has been used to highly-seasoned food, either Soyer, Harvey, or Worcestershire sauces may be used sparingly. Many prefer to season with a greater variety of spices and condiments. I often season with allspice; but camp chowder should be simple, and composed of edibles easily obtainable. Clam chowder is made in the same manner."

Fish-House Punch.

One-quarter of a pint of lemon juice, one-quarter of a pound of white sugar, and two pints and a half of water. One-quarter of a pint of peach brandy; the same of Jamaica rum, and a half pint of cognac; the three latter ingredients mixed separately.

Pineapple Punch.

One slice of pineapple which has stood a day covered with sugar, two bottles of port, one bottle of champagne, and plenty of ice.

Porto Rico Punch.

Black tea and Porto Rico rum, mixed half and half, and sufficient sugar, lemon-peel, and ice.

Nondescript Punch.

One bottle of claret, three-fourths of a tumbler of brandy, a claret glass of Jamaica rum, one bottle of champagne, ice and sugar.

Arrack Punch.

Eight tumblers of Jamaica rum, one and a half of arrack, and one of lemon juice, which together with the rind of three lemons, is to be allowed to stand for ten minutes, when sugar is to be added, and water to twice the amount of the liquor.

Champagne Punch.

One bottle of brandy, one of Jamaica rum, and one of arrack; three and a half pounds of sugar, but no water, four lemons and twelve oranges cut in slices, a large lump of ice. Add champagne to suit the taste immediately before drinking.

Regal Punch.

Peel twenty-four lemons; steep the rinds for twelve hours in two quarts of Jamaica rum. Squeeze the lemons on three and a half pounds of loaf sugar; add two quarts of dark brandy and six quarts of water. Mix all together; add two quarts of boiled milk, stir until the mixture curdles, strain it through a jelly-bag until clear; bottle and cork.

This I have not tried, but give it on good authority.

FRANK FORESTER'S PUNCH.

The rind of a dozen lemons, two tumblerfuls of finely powdered sugar, three pints of pale cognac, two quarts of cold, strong, green tea, strained clear, two flasks of Curaçao, abundance of ice, and a half dozen of champagne. This is an admirable liquor, even without the champagne.

VENISON STEW.

Make a sauce by melting a lump of butter with two mustard-spoonfuls of mustard, two table-spoon fuls of mushroom catsup, and one of sauce, mango sauce being the best; add the juice of half a lemon, one wine glass of sherry, and one of claret. Heat the mixture as hot as possible, and rub in two table-spoonfuls of currant jelly till the whole is perfectly smooth; then take the venison cut in steaks, and previously either roasted or broiled, and warm it thoroughly in the sauce to which the juice of the meat, if any, has been added. Cold meat is redeemed by this process.

And now my friends, if you are ever fortunate enough to have the Superior Fishing I have described, or if the author's good-will may avail even better, and, after the delight and triumph of success, the well-earned prize is brought up properly upon the table, either in the rough woods or the elegant dining-room, and is flanked by such appropriate dishes as circumstances permit, and laid to rest in the best liquor that can be obtained; then your mind, filled with present complacency, must travel back over these pages, and forgetting the faults and pardoning the errors, acknowledge that if in them you have not found an instructor, you have found a brother sportsman; and, for the sake of the bond that binds all members of the gentle craft together, if you cannot conscientiously praise the manner or the matter of these pages, you will utter no word to discourage an effort that, while pointing out and dwelling upon the beauties of nature in our wonderful country, and the pure attractions it offers to the lovers of our art, has principally been to maintain the healthy and ennobling nature of field-sports; to urge the protection, at proper seasons, of the game that still lingers in our woods and waters; and to elevate to a proud standard of honorable, generous, and merciful rivalry the sportsmanship of America.

THE END.

BOOKS

Published by

CARLETON

413 Broad-Way

New-York

1865.

" There is a kind of physiognomy in the titles
of books no less than in the faces of
men, by which a skilful observer
will know as well what to expect from the one as the
other."—BUTLER.

NEW BOOKS

And New Editions Recently Issued by

CARLETON, PUBLISHER,

NEW YORK.

413 *BROADWAY, CORNER OF LISPENARD STREET.*

N.B.—THE PUBLISHER, upon receipt of the price in advance, will send any of the following Books, by mail, POSTAGE FREE, to any part of the United States. This convenient and very safe mode may be adopted when the neighboring Booksellers are not supplied with the desired work. State name and address in full.

Victor Hugo.

LES MISERABLES.—*The best edition,* two elegant 8vo. vols., beautifully bound in cloth, $5.50 ; half calf, . . $10.00

LES MISERABLES.—*The popular edition,* one large octavo volume, paper covers, $2.00 ; cloth bound, . . $2.50

LES MISERABLES.—Original edition in five vols.—Fantine—Cosette—Marius—Denis—Valjean. 8vo. cloth, . $1.25

LES MISERABLES—In the Spanish language. Fine 8vo. edition, two vols., paper covers, $4.00 ; or cloth, bound, . $5.00

THE LIFE OF VICTOR HUGO.—By himself. 8vo. cloth, $1.75

By the Author of "Rutledge."

RUTLEDGE.—A deeply interesting novel. 12mo. cloth, $1.75

THE SUTHERLANDS.— do. . . do. $1.75

FRANK WARRINGTON.— do. . . do. $1.75

LOUIE'S LAST TERM AT ST. MARY'S.— . . do. $1.75

ST. PHILIP'S.—*Just published.* . . do. $1.75

Hand-Books of Good Society.

THE HABITS OF GOOD SOCIETY; with Thoughts, Hints, and Anecdotes, concerning nice points of taste, good manners and the art of making oneself agreeable. Reprinted from the London Edition. The best and most entertaining work of the kind ever published. . . 12mo. cloth, $1.75

THE ART OF CONVERSATION.—With directions for self-culture A sensible and instructive work, that ought to be in th hands of every one who wishes to be either an agreeable talker or listener. . . . 12mo. cloth, $1.50

Miss Augusta J. Evans.

BEULAH.—A novel of great power. . 12mo. cloth, $1.75

Mrs. Mary J. Holmes' Works.

DARKNESS AND DAYLIGHT.— *Just published.*   12mo. cl. $1.50
'LENA RIVERS.—   . . A Novel.   do. $1.50
TEMPEST AND SUNSHINE.—   .   do.   do. $1.50
MARIAN GREY —   .   do   do. $1.50
MEADOW BROOK.—   .   .   .   do.   do. $1.50
ENGLISH ORPHANS.—   .   .   do.   do. $1.50
DORA DEANE.—   .   .   .   do.   do. $1.50
COUSIN MAUDE.—   .   .   . do.   do. $1.50
HOMESTEAD ON THE HILLSIDE.—   do.   do. $1.50
HUGH WORTHINGTON.— *Just published.*   do. $1.50

Artemus Ward.

HIS BOOK.—An irresistibly funny volume of writings by the
immortal American humorist. .   . 12mo. cloth, $1.50
A NEW BOOK.—*In press.*   .   .   .   do. $1.50

Miss Muloch.

JOHN HALIFAX.—A novel. With illust. 12mo., cloth, $1.75
A LIFE FOR A LIFE.—   .   do.   .   do. $1.75

Charlotte Bronte (Currer Bell).

JANE EYRE.—A novel. With illustration. 12mo. cloth, $1.75
THE PROFESSOR.—do.   .   do.   .   do. $1.75
SHIRLEY.—   . do.   .   do.   .   do. $1.75
VILLETTE.—   . do.   .   do.   .   do. $1.75

Edmund Kirke.

AMONG THE PINES.—A Southern sketch. 12mo. cloth, $1.50
MY SOUTHERN FRIENDS.—   do.   do.   . $1.50
DOWN IN TENNESSEE.—Just published.   .   do. $1.5c

Cuthbert Bede.

VERDANT GREEN.—A rollicking, humorous novel of English
student life; with 200 comic illustrations. 12mo. cloth, $1.50
NEARER AND DEARER.—A novel, illustrated. 12mo. clo. $1.50

Richard B. Kimball.

WAS HE SUCCESSFUL?—   A novel.   12mo. cloth, $1.75
UNDERCURRENTS.—   do.   do. $1.75
SAINT LEGER.—   do.   do. $1.75
ROMANCE OF STUDENT LIFE.— do.   do. $1.75
IN THE TROPICS.—Edited by R. B. Kimball.   do. $1.75

Epes Sargent.

PECULIAR.—One of the most remarkable and successful novels
published in this country.   .   . 12mo. cloth, $1.75

A. S. Roe's Works.

A LONG LOOK AHEAD.— A novel. 12mo. cloth, $1.50
TO LOVE AND TO BE LOVED.— do. . . do. $1.50
TIME AND TIDE.— do. . . do. $1.50
I'VE BEEN THINKING.— do. . . do. $1.50
THE STAR AND THE CLOUD.— do. . . do. $1.5c
TRUE TO THE LAST.— do. . . do. $1.50
HOW COULD HE HELP IT.— do. . . do. $1.50
LIKE AND UNLIKE.— do. . . do. $1.50
LOOKING AROUND.— *Just published.* do. $1.50

Walter Barrett, Clerk.

OLD MERCHANTS OF NEW YORK.—Being personal incidents, interesting sketches, bits of biography, and gossipy events in the life of nearly every leading merchant in New York City. Three series. . . 12mo. cloth, each, $1.75

T. S. Arthur's New Works.

LIGHT ON SHADOWED PATHS.—A novel. 12mo. cloth, $1.50
OUT IN THE WORLD.— do. . do. $1.50
NOTHING BUT MONEY.— do. . do. $1.50
WHAT CAME AFTERWARDS.—*In press.* . do. $1.50

Orpheus C. Kerr.

ORPHEUS C. KERR PAPERS.—Three series. 12mo. cloth, $1.50
THE PALACE BEAUTIFUL.—And other poems. do. $1.5c

M. Michelet's Works.

LOVE (L'AMOUR).—From the French. 12mo. cloth, $1.50
WOMAN (LA FEMME.)— do. . . do. $1.50

Novels by Ruffini.

DR. ANTONIO.—A love story of Italy. 12mo. cloth, $1.75
LAVINIA; OR, THE ITALIAN ARTIST.— do. $1.75
VINCENZO; OR, SUNKEN ROCKS.— 8vo. cloth, $1.75

Rev John Cumming, D.D., of London.

THE GREAT TRIBULATION.—Two series. 12mo. cloth, $1.50
THE GREAT PREPARATION.— do. . do. $1.50
THE GREAT CONSUMMATION.— do. . do. $1.50

Ernest Renan.

THE LIFE OF JESUS.—Translated by C. E. Wilbour from the celebrated French work. . . 12mo. cloth, $1.75
RELIGIOUS HISTORY AND CRITICISM.— 8vo. cloth, $2.50

Cuyler Pine.

MARY BRANDEGEE.—An American novel. . ⌐ $1.75
A NEW NOVEL.—*In press.* $1.75

Charles Reade.
THE CLOISTER AND THE HEARTH.—A magnificent new novel, by the author of "Hard Cash," etc.   .   8vo. cloth, $2.00

The Opera.
TALES FROM THE OPERAS.—A collection of clever stories, based upon the plots of all the famous operas.   12mo. cl., $1.50

J. C. Jeaffreson.
A BOOK ABOUT DOCTORS.—An exceedingly humorous and entertaining volume of sketches, stories, and facts, about famous physicians and surgeons.   12mo. cloth, $1.75

Fred. S. Cozzens.
THE SPARROWGRASS PAPERS—A capital humorous work, with illustrations by Darley.   .   .   12mo. cloth, $1.50

F. D. Guerrazzi.
BEATRICE CENCI.—A great historical novel. Translated from the Italian; with a portrait of the Cenci, from Guido's famous picture in Rome.   .   .   12mo. cloth, $1.75

Private Miles O'Reilly.
HIS BOOK.—Comic songs, speeches, &c.   12mo. cloth, $1.50
A NEW NOVEL.—*In press.*   .   .   .   do. $1.50

The New York Central Park.
A SUPERB GIFT BOOK.—The Central Park pleasantly described, and magnificently embellished with more than 50 exquisite photographs of the principal views and objects of interest. A large quarto volume, sumptuously bound in Turkey morocco, ... $30.00

Joseph Rodman Drake.
THE CULPRIT FAY.—The most charming faery poem in the English language. Beautifully printed. 12mo. cloth, 75 cts.

Mother Goose for Grown Folks.
HUMOROUS RHYMES for grown people; based upon the famous "Mother Goose Melodies."   .   .   12mo. cloth, $1.00

Mrs. ——— ———
FAIRY FINGERS.—A new novel.   .   12mo. cloth, $1.75
THE MUTE SINGER.— do.   *In press.*   do. $1.75

Robert B. Roosevelt.
THE GAME FISH OF THE NORTH.—Illustrated. 12mo. cl. $2.00
SUPERIOR FISHING.—*Just published.* do.   do. $2.00
THE GAME BIRDS OF THE NORTH.—*In press.*   .   . $2.00

John Phoenix.
THE SQUIBOB PAPERS.—With comic illustr. 12mo. cl., $1.50

N. H. Chamberlain.
THE AUTOBIOGRAPHY OF A NEW ENGLAND FARM-HOUSE.—$1.75

Amelia B. Edwards.
BALLADS.—By author of "Barbara's History." $1.50

S. M. Johnson.
FREE GOVERNMENT IN ENGLAND AND AMERICA.—8vo. cl. $3.00

Captain Semmes.
THE ALABAMA AND SUMTER.— . . 12mo. cl. $2.00

Hewes Gordon.
LOVERS AND THINKERS.—A new novel. . . . $1.50

Caroline May.
POEMS.—*Just published.* . . . 12mo. cloth, $1.50

Slavery.
THE SUPPRESSED BOOK ABOUT SLAVERY.—12mo. cloth, $2.00

Railroad and Insurance
ALMANAC FOR 1865.—Full of Statistics. . 8vo. cloth, $2.00

Stephen Massett.
DRIFTING ABOUT.—Comic book, illustrated. 12mo. cloth, $1.50

Thomas Bailey Aldrich.
BABIE BELL, AND OTHER POEMS.—Blue and gold binding, $1.50
OUT OF HIS HEAD.—A new romance. 12mo. cloth, $1.50

Richard H. Stoddard.
THE KING'S BELL.—A new poem. . 12mo. cloth, 75 cts.
THE MORGESONS.—A novel. By Mrs. R. H. Stoddard. $1.50

Edmund C. Stedman.
ALICE OF MONMOUTH.—A new poem. 12mo. cloth, $1.25
LYRICS AND IDYLS.— . . . . do. $1.25

M. T. Walworth.
LULU.—A new novel. . . . 12mo. cloth, $1.50
HOTSPUR.— do. . . . . . do. $1.50

Author of "Olie."
NEPENTHE.—A new novel. . . 12mo. cloth, $1.50
TOGETHER.— do. . . do. $1.50

Quest.
A NEW ROMANCE.— . . . 12mo. cloth, $1.50

Victoire.
A NEW NOVEL— . . . 12mo. cloth, $1.75

James H. Hackett.
NOTES AND COMMENTS ON SHAKSPEARE.— 12mo. cloth, $1.50

Miscellaneous Works.

JOHN GUILDERSTRING'S SIN.—A novel. . 12mo. cloth, $1.50
CENTEOLA.—By author "Green Mountain Boys." do. $1.50
RED TAPE AND PIGEON-HOLE GENERALS.—.   do. $1.50
THE PARTISAN LEADER.—By Beverly Tucker.   do. $1.50
ADAM GUROWSKI.—Washington diary for 1863.   do. $1.50
TREATISE ON DEAFNESS.—By Dr. E. B. Lighthill. do. $1.50
THE PRISONER OF STATE.—By D. A. Mahoney.   do. $1.50
AROUND THE PYRAMIDS.—By Gen. Aaron Ward. do. $1.50
CHINA AND THE CHINESE.—By W. L. G. Smith. do. $1.50
THE WINTHROPS.—A novel by J. R. Beckwith.   do. $1.75
SPREES AND SPLASHES.—By Henry Morford.   do. $1.50
GARRET VAN HORN.—A novel by J. S. Sauzade.   do. $1.50
SCHOOL FOR THE SOLDIER.—By Capt. Van Ness. do.   50 cts.
THE YACHTMAN'S PRIMER.—By T. R. Warren.   do.   50 cts.
EDGAR POE AND HIS CRITICS.—By Mrs. Whitman. do. $1.00
ERIC; OR, LITTLE BY LITTLE.—By F. W. Farrar. do. $1.50
SAINT WINIFRED'S.—By the author of "Eric."   do. $1.50
A WOMAN'S THOUGHTS ABOUT WOMEN—   .   do. $1.50
THE SEA.—By Michelet, author of "Love."   do. $1.50
MARRIED OFF.—Illustrated satirical poem.   .   do.   50 cts.
SCHOOL-DAYS OF EMINENT MEN.—By Timbs.   do. $1.50
ROMANCE OF A POOR YOUNG MAN.—.   .   do. $1.50
THE FLYING DUTCHMAN.—J. G. Saxe, illustrated. do.   75 cts.
ALEXANDER VON HUMBOLDT.—Life and travels.   do. $1.50
LIFE OF HUGH MILLER—The celebrated geologist. do. $1.50
LYRICS OF A DAY—or, newspaper poetry. .   do. $1.00
THE U. S. TAX LAW.—"Government Edition."   do. $1.00
TACTICS; or, Cupid in Shoulder-Straps.   .   do. $1.50
DEBT AND GRACE.—By Rev. C. F. Hudson.   do. $1.75
THE RUSSIAN BALL.—Illustrated satirical poem. do.   50 cts.
THE SNOBLACE BALL.— do.   do. do.   do.   50 cts.
THE CHURCH IN THE ARMY.—By Dr. Scott.   do. $1.75
TEACH US TO PRAY.—By Dr. Cumming. .   do. $1.50
AN ANSWER TO HUGH MILLER.—By T. A. Davies. do. $1.50
COSMOGONY.—By Thomas A. Davies. .   8vo. cloth, $2.00
TWENTY YEARS around the World. J. Guy Vassar. do. $3.75
THE SLAVE POWER.—By J. E. Cairnes. .   .   do. $2.00
RURAL ARCHITECTURE.—By M. Field, illustrated. do. $2.00

www.ingramcontent.com/pod-product-compliance
Lightning Source LLC
Chambersburg PA
CBHW031401270326
41929CB00010BA/1272